Owning Scientific and Technical Information

Regards to Larry during a stimulating Sabbatical

Y0-CDH-352

Owning Scientific and Technical Information

Value and Ethical Issues

Edited by

Vivian Weil

and

John W. Snapper

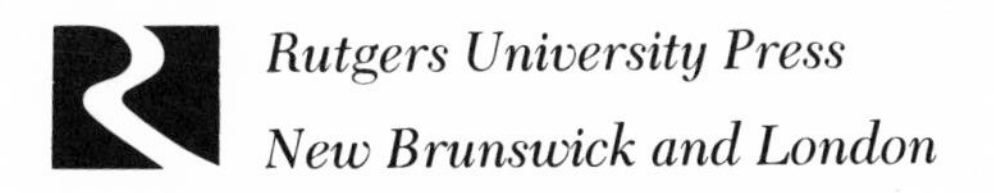

Rutgers University Press

New Brunswick and London

Copyright © 1989 by Rutgers, The State University
All Rights Reserved
Manufactured in the United States of America

Library of Congress Cataloging-in-Publication Data

Owning scientific and technical information : value and ethical issues / [edited by] Vivian Weil, John W. Snapper.
p. cm.
Bibliography: p.
Includes index.
ISBN 0-8135-1454-1 (cloth) ISBN 0-8135-1455-X (pbk.)
1. Intellectual property—United States—Moral and ethical aspects. 2. Patent laws and legislation—United States—Moral and ethical aspects. I. Weil, Vivian. II. Snapper, John W.
KF2979.A2096 1990
346.73 04'8—dc19
[347.30648] 89-30377
CIP

British Cataloging-in-Publication information available

The following chapters appeared, in a slightly different form, in *Science, Technology & Human Values* 12 (Winter 1987). Copyright © 1987 by Sage Publications, Inc.; reprinted by permission of the publisher.
"Introduction" by Vivian Weil
Chapter 4, "Ethical Issues in Proprietary Restrictions on Research Results," by Alan H. Goldman
Chapter 5, "Patenting and Academic Research: Historical Case Studies," by Charles Weiner
Chapter 6, "Biotechnology, Plant Breeding, and Intellectual Property: Social and Ethical Dimensions," by Frederick H. Buttel and Jill Belsky
Chapter 8, "Innovation and Competition: Conflicts over Intellectual Property Rights in New Technologies," by Pamela Samuelson
Chapter 13, "Commentary: Legal and Ethical Issues," by Gerald Dworkin

Contents

Figures and Tables

Figures

Tables

Acknowledgments

The editors are grateful to many people for their support and help in bringing this volume to completion. The contributions of three people deserve special acknowledgment. Rochelle Hollander, Program Manager at the National Science Foundation, gave valuable criticism of the initial conception of the study and continued to give encouragement as the collection of papers took shape. Karen Reeds of Rutgers University Press provided the kind of supportive guidance that editors wish for and are rarely lucky enough to get. Rebecca Newton, Secretary to the Center for the Study of Ethics in the Professions, made possible the preparation of the manuscript by her patience, perseverance, and great skill.

Initial research for this volume was supported by the National Science Foundation project #RII-830-9873, "Ethical Implications of Trade Secrets, Patents, and Related Property Controls for Science and Technology." The reviewers on the project's Advisory Panel gave substantive advice and suggestions that enhanced both the scheme and the content of this volume. They include Lawrence Becker, Jules Coleman, and Harold Relyea. We are pleased to acknowledge their help and, of course, absolve them of any deficiencies that remain. We also benefited from the comments of the reviewers selected by Rutgers University Press, Linda Garcia and Judith Lichtenberg.

We wish to thank the participants in a National Science Foundation conference that was part of this project. A number of the authors in this volume attended that meeting. Other invited participants were Roger Beachy, Robert Benson, David Blumenthal, Myles Boylan, Anne Branscomb, Howard Bremer, Lewis Collens, Joseph Fried, Stephen J. Gage, Jorge Goldstein, Jack Golodner, George Indig, Morton Kamien, Michael Keplinger, Jack Kloppenburg, Jr., Orin Laney, Stanley Liebowitz, James Nickel, John Palmer, Richard Rozek, Dorothy Shrader, James Stukel, and William Thomas.

Ancestors of five of the papers, those by Frederick H. Buttel and Jill Belsky, Gerald Dworkin, Alan H. Goldman, Pamela Samuelson, and Charles Weiner, and material in the introduction written by Vivian Weil appeared in *Science, Technology and Human Values* 12 (January 1987). We are grateful to Marcel La Follette, former editor of *Science, Technology and Human Values*, for her editorial contributions and for her encouragement.

We wish to thank the contributors to this volume who ventured into cross-disciplinary study that often drew them outside their professionally defined specialties. We also gratefully acknowledge the support of the National

Science Foundation. The authors whose work was supported by the NSF grant are Frederick H. Buttel, Patrick Croskery, Duncan M. Davidson, Rochelle Cooper Dreyfuss, Gerald Dworkin, Alan H. Goldman, Patrick D. Kelly, Arthur Kuflik, A. J. Lemin, Pamela Samuelson, Charles Weiner, and Sidney G. Winter. The views expressed in the volume are those of the authors of the respective essays.

Vivian Weil
John W. Snapper
Chicago, Illinois
December 1988

Introduction

Our intellectual property system has jumped into prominence in this decade. Discussion about the scope of patents, copyrights, and related property controls engages a much broader audience than previously. The ranks of those with a direct stake in our intellectual property system have increased sharply.

There is no mystery about the causes of this heightened concern. As new technologies have become the basis of new economically critical industries and transformed older industries, the technologies have become the focus of debates over proprietary protection. However, because our innovative technology includes novel features, it is not easily protected by established policy. The new sorts of "products" and "processes" created by the biotechnology and software industries, for instance, challenge traditional classifications of patent and copyright law. Moreover, such items as plant varieties, which were freely available early in the century, have been turned into commodities by extension of the patent law to cover these "inventions." Questions about whether the agribusiness industry needs or deserves special protections and whether the protections have the desired effects have only intensified with the latest expansion of the patent law to cover virtually all plant varieties.

In addition, proprietary protection has moved into new research settings. Universities, for instance, no longer keep their distance from the intellectual property system. As university-based researchers take advantage of new opportunities for commercial gain, they precipitate worries that established academic ideals and practices may be undermined.

Such changes have given rise to a number of important controversies with ethical implications. Examples include debates about how to protect software without closing off lines of innovation or how to satisfy patenting requirements for microorganisms while maintaining cell-line exchange practices. The resolution of these disputes has significance far beyond the concerns of the parties in conflict. Our intellectual property arrangements really matter; they affect the ordinary lives of nonscientists and nontechnologists in profound ways. The opportunities for ownership of new technologies help to determine which technologies are developed and which die on the drawing boards. Some fail to develop as we might wish because of obstacles to protection or the absence of such obstacles. The technologies in question may be vitally related to health, agricultural productivity, manufacturing, or information and communications.

The controversies over recent changes deal with very deep and serious questions for society. How do we best stimulate and support fundamental scientific research? What is the impact of specific proprietary controls on scientific research and the development of technology? How do we devise or modify proprietary controls for optimum economic effects? What is owed to innovators and creators of new works and how can they be rewarded without unduly interfering with competition and the dissemination of new findings and technologies? How can proprietary controls be meshed with traditional patterns of free exchange in science and in the universities? These are questions that lawyers, legislators, and economists have addressed at the practical level. Philosophers, legal theorists, and economic theorists have concentrated upon the underlying abstract justifications. So far as we know, these approaches have not heretofore been brought together.

The task of thinking through, in a comprehensive fashion, the foundations and justifications of intellectual property controls and assessing effects of various arrangements on research and development becomes ever more urgent as increasing strain is placed on the system. Such ethical issues as just rewards for innovators, fairness in the allocation of benefits, public access to new technology and knowledge, and quality control over innovations are embedded in economically complex circumstances. Empirical data about these circumstances are scarce and difficult to obtain. Nevertheless, we need to replace presumptions about effects with information about how arrangements actually work out, and we have to bring theoretical considerations to the test in actual circumstances. To accomplish this task, a multidisciplinary approach is necessary.

The papers in this volume represent such a response to the task of investigating the foundations of our intellectual property system. Theoretical questions raised from the perspectives of law, philosophy, and economics are explored in conjunction with examination of the actual functioning of the intellectual property system. A broad framework for discussion is provided in an initial overview paper by a legal theorist. Then authors representing different disciplines investigate the changing conditions of scientific research and technological development from their distinctive perspectives. Case studies of current and historical instances establish a context for the assessment of proprietary protection. These cases provide evidence about the actual workings of intellectual property policies, and a focus for the rationales and theoretical critiques. They exhibit the issues at stake in the subsequent debate over presumptions and justifications. The final section focuses once more on underlying assumptions about rights and results, now against the background of the examination of cases. The volume concludes with a philosophical, analytical approach to the literature of this interdisciplinary area and a carefully selected annotated bibliography. The approach to the literature provides a comprehensive structure for continuing the discussion.

The present volume concentrates on issues relating to innovative research in the environment created by intellectual property policy. Hence such disputes as those about artists' rights, colorization of movies, and the extent of permissible coping of television shows, videotapes, and live concerts are outside the focus of this volume. Although these other issues are both important to our lives and filled with interesting controversy, they are passed over here so as to permit a more concentrated attack on issues relating to scientific and technological research.

Ownership Controls

There can be no simple definition of *intellectual property*. The intuitive notion is that intellectual property arrangements give owners control over things that are intellectually understood and appreciated. Examples include ownership claims over new machines, poems, and methods used to synthesize a chemical. But these property arrangements exhibit such variety that it is very hard to say what it is about them that makes them all "proprietary arrangements." The difficulty is not special to the notion of intellectual property in particular, but to the broader notion of property in all its applications. A bit of reflection will show that we have no clear intuitions on what it means to say in general that we "own" an item.

An owner of an item has some say over who may use the item and how. There is, however, no single form of control that always holds in all ownership claims. For instance, the owner of a house typically may remodel it (a right to alter the item), sell it (a right to transfer), decide how to use the house (a right to use), and much more. But if the house has landmark status, the right to alter may be severely limited. If the house has been leased, the right to sell is limited. If the house is in a zoned area of the city, its use is limited. We can easily give examples where the owner of a house has very few of the rights usually associated with ownership of a building.

The notion of property is further complicated by the variety of very different sorts of things that we claim to own. We may own a plot of land, or the trees on the land, or the fruit that we expect the trees to bear, or access to the space above the land. We may own a corporation including all its real assets, or the name and reputation of a corporation. We may own a book of poetry (that is, the material item consisting of paper and ink), or the poems included in the book and none of their particular inscriptions. Obviously, these very different sorts of things are owned in very different sorts of ways. A car owner decides who may move the car and whether to destroy it in a steel compactor. A plot of land can be neither moved nor destroyed, but the owner decides who may occupy it. A poem can neither be moved nor destroyed nor occupied, but the owner decides where it may be inscribed. Different sorts of things demand different notions of ownership.

Although there is no single form of control over objects that characterizes ownership, philosophers often say that there is a "bundle" of rights that are associated with property relations. Any particular example of an ownership arrangement will usually include more than one of the rights from the bundle. There are differences of opinion on what rights comprise the bundle of rights that typify property relations. But the following rights are commonly included: (1) a right to destroy the item; (2) a right to say who may use the item and how; (3) a right to income generated by the item or its use; (4) a right to transfer the item by sale, inheritance, or gift; (5) a right to modify the item; (6) a right to determine who may physically touch, occupy, or move the item; and (7) a right to enjoy or to benefit from the use of the item. Many writers would add a list of responsibilities attached to ownership claims. These include the owner's liability for harm due to dangers created by the property and responsibility for upkeep on the property. Although there are certainly additional rights associated with property that would have to be discussed in a broad theoretical study of property, this list indicates the sorts of relations and controls that are at issue in a dispute over the justification of property.

What Is Owned

Property is generally classified as tangible (for example, cars, gold, and plots of land) or intangible (for example, poems, names of corporations, and industrial processes). Intellectual property is an example of the latter. Although this broad distinction ignores many important differences between varieties of proprietary arrangements within the classifications, it draws attention to the fact that the notion of intellectual property differs from the everyday notions of property that apply to our material possessions. And the familiar notions of material possession (with their varied bundles of rights) may not be much of a guide to how we are to define property in this realm of untouchable, unmovable, undestroyable, unoccupiable entities. We have a tremendous amount of leeway in how we create proprietary arrangements in this area. Consider, for instance, the term of ownership. Although there is usually no limit on how long we may own a material object, patents on inventions are usually limited to seventeen years and copyrights on manuscripts to about seventy-five years.

Legal Protections

The particular bundles of rights associated with the various forms of intellectual property are grounded in several different legal traditions and supported by several different forms of authority. We may roughly outline the more popular forms of legal protection for intellectual property in the United States as follows:

1. Federally established licensing procedures
   a. Patents
   b. Copyrights
   c. Trademarks
   d. Generic protections
2. Trade practices and agreements
   a. Trade secrets
   b. Contracts
   c. Other arrangements

Given the complexity of our intellectual property system, simple pictures must be misleading, and the picture suggested by this outline is no exception. The outline suggests, for instance, that trade practices are not established by federal policy. But trade practices are regulated by, among other things, federal antitrust law. Moreover antitrust law places limitations on patent licenses. So even though this outline lists them separately, some constraints on trade practices should be discussed under the category of patents. All the same, we can use this outline as a guide for an overview of the complex of laws that define intellectual property.

Under the U.S. Constitution, Congress has the power to grant inventors and authors exclusive rights over their works. This is the basis for patent and copyright protections. For both forms of protection, the Constitution says that the purpose of the law is to "promote the progress of science and the useful arts." In theory, it would be possible to challenge a clause in the patent law on the grounds that it hinders progress in science. And many clauses of the patent law can be viewed as a natural consequence of its constitutional purpose. For instance, patents are not granted on inventions that have been developed and withheld from the public for an extended period, because new scientific progress is not promoted by protections on established technology.

Patents

Patents traditionally are divided into three categories: inventions (such as machines), processes (such as a way to synthesize a chemical), and substances (such as a new form of plastic). In recent years, patents have been extended to cover such things as new plant varieties. The patent holder has the right to determine who may produce, use, or sell his invention, process, or substance. In all cases, the protection is only over material objects based on the inventor's design or material implementations of the inventor's process, and never over the theoretical basis for the innovation. So the inventor of a car may determine who builds the cars, but not who may publish descriptions of the new technology that goes into the car or who may develop that technology in yet newer inventions. If an inventor discovers that his or her rights are infringed, he or she may sue for damages and a halt to the infringing practice.

Copyrights

Copyrights protect ways of expressing ideas. As the name suggests, the copyright holder has a right to determine who may copy his words. That right does not extend to who may discuss the ideas expressed by those words. For instance, Einstein may copyright his essay on the theory of relativity, but he may not prevent others from researching his theory or presenting it in their own words. When there is only one standard way to present an idea (for instance, in mathematical notation), then control over that form of expression would amount to control over all discussion of the idea and, hence, no copyright protection is available. In recent years, copyright protections have been extended to include pictures, computer programs, and design elements of products. Many of these extensions are very controversial.

Trademarks

A trademark is literally a mark, typically stamped on a product, to indicate its producer. The producer has control only over use of the mark, not over any features of the product. A clothing designer's trademark signature on the back pocket of his or her distinctive jeans does not protect him or her from someone who copies those jeans exactly without the trademark. Although a high technology firm may place a trademark on its innovative products as a sign that it is up to its standards, these marks are really meant to protect the business relationship and are only incidentally indications of technological excellence. Although viewed as the third member of the patent-copyright-trademark trio (all supervised by one government commission), trademark control is based on Congress's right to regulate interstate commerce rather than its power to promote scientific progress. Trademarks are important for the protection of commerce but are tangential to most of the intellectual property concerns discussed in this book.

Generic Protections

Special forms of protection are sometimes provided for particular industries whose products do not neatly satisfy criteria for patents and copyrights. There is, for instance, much debate over the possibility of a new form of protection for computer software, which is sufficiently different from both patentable machines and copyrightable manuscripts that it is only partially protected by existing law. In response to special needs, Congress has occasionally enacted laws that protect only one distinctive sort of innovation. Two examples of such special *sui generis* protections discussed in this book are the Semiconductor Chip Protection Act and the Plant Variety Protection Act.

While the protections discussed above are fairly well defined in federal statutes and legal decisions associated with those statutes, the remaining trade practices that create protections for intellectual property form a confused cluster with no single, well-defined legal basis. Here we run into a jumble of proprietary notions, based on common business traditions, supported by common law traditions, and regulated in different (often conflicting) ways by state and federal agencies. To a large extent, the underlying notion is that some activities are unfair trade practices. A classic example is the instance of the businessperson who learns a competitor's plans by bribing a private secretary. In that case, the victim might (depending on circumstance and state law) be able to enjoin the briber from making use of the knowledge or even be recompensed for damages. For our concerns here, we must take note that it is unfair to misappropriate what is commonly recognized as intellectual property.

Trade Secrecy

Trade secrecy is defined in the *Restatement of Torts* in terms of the effort that the owner takes to keep it secret (it is not secret if you discuss it over lunch) and the effort that others would need to discover the secret independently (it is not secret if any engineer could work it out). The *Restatement* then notes that it is unfair to discover those secrets by "means which fall below the generally accepted standards of commercial morality and reasonable conduct." Typical means of unfair discovery are disclosure by disloyal employees and industrial espionage.

The *Restatement* is not a document with formal legal authority. It is something like a legal dictionary, created by a prestigious committee of lawyers, and given considerable weight in legal dispute. Very similar definitions of trade secrecy are recognized by statute in several states. If a secret is lost in spite of a reasonable attempt by the owner to maintain secrecy, then the courts may uphold the owner's rights to control the information through injunctions against unauthorized use. This establishes a form of legal ownership that goes far beyond the underlying right to refuse disclosure. Trade secrecy law thus creates a a type of ownership over intellectual property akin to the more explicit ownership relations established in patent and copyright law.

In some contexts, trade secrecy complements patent protections. Patents are not granted on innovations that are obvious or straightforward developments of public information. Thus, an inventor could forfeit the right to patent if the work is not kept a trade secret while being developed and tested to the point that it can be patented. In other contexts, trade secrets are alternatives to patents. A process for the synthesis of chemicals, for instance, may be commercially exploited without disclosing it in a patent application. If the inventor takes this approach and continues to keep the process secret after it has been developed and tested, the process becomes an established industrial practice and the right to patent is forfeited. Trade secrecy is often preferred by industries, such as the software industry, that find it difficult to derive benefits from use of patentlike protections.

Contracts

Contractual relationships can be used to establish proprietary protections. For instance, an inventor may make results available for commercial use only to users who agree to pay royalties, keep the results secret, and go back to the original inventor for further developments of the technology. Such contracts give the inventor ownership rights, at least among the parties to the contract. Some aspects of these contracts are viewed as trade secrecy arrangements. But there is great difference between violation of a contract to keep a secret and industrial espionage where there is no contract between disputants.

We should not think of contractual agreements as simply private arrangements without governmental endorsement. When a contract is violated, it may be enforced with government authority in courts that abide by the law of the land. The courts would, for instance, usually uphold a contract wherein a user agrees to go back to the inventor for modifications on a machine. But the courts might not uphold an agreement that demands that the user go back to the inventor for all services and additional developments in related technology. The latter could give the inventor too much monopolistic control over the technology. Since the definition of an enforceable contract is a public matter, the nature of these proprietary arrangements is in part determined by public policy.

Other Arrangements

There are many additional arrangements under which items might be viewed as intellectual property. We cannot hope to capture them all in our rough outline. Plagiarism, for instance, may be viewed as the violation of a notion of ownership accepted by tradition within the scientific community. Any scientist who presents a theory discovered by another is expected to acknowledge the original work. Failure to do so is condemned as plagiarism and can lead to exclusion from the scientific community. A right to recognition for their work and freedom from plagiarism is what scientists and academics usually have in mind when they refer to "so-and-so's theory," as if so-and-so owned it. But since plagiarism of a theory need not involve a copyright infringement, this notion of ownership is largely due to an extralegal tradition that differs from any discussed above. Additional notions of intellectual property are established or influenced by international treaty, employment conditions, inheritance laws, and so forth.

The Normative Study of Intellectual Property

The essays in this book emphasize normative rather than descriptive analysis of intellectual property. That is, they are less concerned with describing the bundles of rights that *are* granted to owners of intellectual property than with determining what rights *should* be granted to owners of intellectual property. These normative studies seek ways to justify or to choose between alternative systems of rights. We should note that there are several levels at which these studies may take place.

Consider, for instance, the dispute over whether copyrights should be granted on published computer programs that can only be understood by a computer. Traditionally, a copyright applicant publicly files a readable form of the entire manuscript to be copyrighted. The computer industry is now permitted an exception to this rule and granted copyrights on programs that are largely kept in unreadable machine format. Assuming that the intent of Congress in granting copyrights on programs should be to promote progress in the computing arts, we may justify these special arrangements with evidence that the new copyright rules are in fact contributing to a climate favoring research and development in the computer industry. This would be a direct validation of the rules in terms of a constitutional assumption about how copyrights are to be justified.

We may, however, look at the issue of copyrights on unpublished programs on a more theoretical level. Still granting that copyrights should promote computing arts, we may seek an economic analysis of how the copyright achieves that end. The copyright system may, for instance, provide an incentive to researchers entering the computer field who hope to make a profit from their work under the new copyright provisions. Alternatively, by financially rewarding holders of present copyrights, the system may result in an allocation of wealth in the computing fields to those whose past record shows that they are most likely to continue to create new computer programs. Both analyses may explain why the copyright system creates a climate favoring research and development in the computer industry. But the two analyses may have different normative consequences. The former, for instance, places a higher premium on public disclosure of copyrighted programs than the latter, since public disclosure is a help to new researchers entering an industry. This is a normative analysis based on a theoretical study of how the copyright system works to achieve the assumed constitutional goal of promoting the programming arts.

On a yet more abstract level, we may question whether pecuniary return for innovative programs should play such a central role in our discussion. Copyrights are also used for other reasons. Programmers use copyright, for instance, as a legal method to ensure that their colleagues acknowledge original work, even when the copyright holder forgoes financial reward. Copyrights on sensitive programs (such as expert medical systems) also permit copyright holders to determine who is using their work and thereby assure themselves that unqualified users are not using their work in dangerous ways. These issues do not bear directly on whether the system promotes additional research and development. We may now seek a normative evaluation of the weight to be given these alternative justifications, balanced against economic arguments concerning the use of copyrights to promote research. This is a normative analysis of alternative justifications offered for the copyright rules.

We may also provide normative critiques of the alternative justifications themselves. We may seek means to justify philosophically our interest in the promotion of the programming arts, in programmers' expectations of recognition for creative work, and in the need to control software that could be misused. To some extent the critiques are based on recognition of potential benefits to society. To some extent they may be founded on a sense of the personal rights of the programmer. This is a normative analysis, on a very deep level, of the assumptions that underlie our more direct justifications of the copyright rules.

It is a mistake to divorce the applied analysis of copyrights on unpublished programs from the more abstract, theoretical issues. Each discussion bears directly on the other. Yet we must constantly guard against the two obvious pitfalls: a refusal to recognize the significance of theoretical studies and a tendency to theorize without understanding the everyday concerns of practicing researchers. Without falling prey to either error, there remains a myriad of ways to balance theory and application. Each study in this volume strikes a balance between theoretical and applied issues in a different manner.

Outline of the Present Study

The normative study of intellectual property in scientific research contexts must be interdisciplinary. Intellectual property is an immediate concern of working scientists, engineers, and inventors whose research practices are in part dictated by the conditions for the commercial exploitation of their results. It is the professional focus of lawyers and legislators working to interpret and refine the system to make it fair to consumers, researchers, and all others with a stake in innovative technology. And it demands the attention of economists, historians, philosophers, social scientists, and others who seek to understand how the system works both in practice and in theory. The present collection attempts to construct a coherent picture of our intellectual property system out of these varied perspectives.

This volume grew out of a day-and-a-half-long interdisciplinary conference presented by the Center for the Study of Ethics in the Professions at the Illinois Institute of Technology in October 1985. A select group of invited participants, representing a broad range of disciplines and perspectives, discussed theoretical analyses and commissioned case studies on the effects of intellectual property on various research industries. The conference included representatives of both the legal and research staffs of large corporations, historians of science and technology, economists, philosophers, representatives of government agencies charged with implementing intellectual property policy, and others with established interests in the issues. Many of the essays in the present volume are descendants of papers that were prepared by participants at the conference. Additional papers were commissioned in areas where the editors saw the need for supplementary studies. The papers look at particular effects of intellectual property on research practice and at theoretical justifications for those intellectual property arrangements.

The papers collected here have been separated into four sections. The first provides an introduction to the varied approaches to the subject. The next two focus on particular controversies—historical and contemporary—that have involved the intellectual property system. The final section returns to theoretical discussion of themes underlying those issues and controversies.

The first section exhibits the range of interests in and alternative approaches to intellectual property. Each paper shows, first, how a particular discipline attempts to deal with issues in intellectual property and, second, what that discipline takes to be the central controversies in the field. Through a survey of contemporary legal disputes, Rochelle Dreyfuss provides a legal framework for the subsequent studies. Sidney Winter shows how economic analyses of intellectual property have oversimplified the issues and explains what must be done to provide a more sophisticated appreciation of how intellectual property promotes commercial research interests. Patrick Kelly looks at the law's response to concerns of practicing researchers. He shows how we can better accommodate the needs of research scientists with slight changes in the legal system. Alan Goldman applies traditional philosophical distinctions drawn from discussions of social justice to analyze popular justifications of our intellectual property system.

The historical case studies grant the wisdom of hindsight in appreciating controversial aspects of the intellectual property system. Charles Weiner concentrates on instances when use of proprietary protections created major controversies for public and privately funded research institutions. Frederick Buttel and Jill Belsky trace the history of complex responses by one industry and its related research institutions to revisions in property arrangements designed to protect that industry.

The contemporary case studies highlight two areas of continued controversy for contemporary technology. Duncan Davidson and Pamela Samuelson disagree over the value of copyright for the protection of computer software. Alan Lemin, Carl Cranor, and Pamela Samuelson try to assess recent attempts to accommodate the interests of biotechnology research with patent protections.

Finally we return to theoretical, philosophical commentaries on the issues. Arthur Kuflik seeks a coherent philosophic approach to the special property interests discussed in the case studies. Michael Davis provides an understanding of power given to Congress, under the Constitution, to grant intellectual property rights. Leonard Boonin summarizes the legal and philosophical basis for proprietary protection and focuses on the tensions in universities between traditional values supporting fundamental research and intellectual property protection. Patrick Croskery provides a structure within which we can appreciate the different directions taken in the study of intellectual property. We then conclude with a select, though broad-based, annotated bibliography of the normative work on intellectual property and the interaction between our intellectual property system and research practices. The bibliography is shaped as a guide for further studies rather than as a compendium of references for this volume. Several essays include additional references to specialized issues.

The Context: Ethics, Society, and Law

PART I

The chapters in this part provide alternative frameworks for the study of research practices in an intellectual property environment, reflecting the approaches of different disciplines. Rochelle Dreyfuss tells us how a lawyer would view the issues. Sidney Winter introduces us to an economic analysis. Patrick Kelly gives an example of the difficulty working scientists face when the criteria for property protection conflict with traditions for scientific recognition. And Alan Goldman puts the issues into the context of traditional philosophy. Each point of view is distinctive, and none should be ignored in a broad study of the issues.

Since intellectual property is to such a large degree defined by the law, we begin with a lawyer's overview of the issues. Dreyfuss brings forth those which seem most pressing for the legal system. She introduces a broad range of intellectual property issues and ties them all to problems confronting scientists and engineers. For instance, the need to balance individual scientists' rights with social benefits bears on a dispute over an attempt by the estate of Elvis Presley to claim ownership over the "Elvis look." The conflict between the ideals of open science and the demands of trade secrecy are reflected in disputes over Union Carbide's attempt to control information concerning the Bhopal disaster.

Dreyfuss demonstrates that decisions on how to protect technological innovation cannot be divorced from other issues in the law. Consider, for instance, a recent proposal to create distinctive forms of intellectual property that protect a single technology. Such "generic protections" are being considered for the computer software industry and have to some extent been enacted for the computer chip industry. These suggestions, however, cannot be divorced from issues—not obviously related to the needs of scientists and engineers—that arise out of international agreements on intellectual property law. One basic point is that new forms of generic protection must be internationally recognized if they are to be effective in the modern computer industry. Since it is difficult to establish international agreements, there is great incentive to accommodate the computer industry within traditional forms of protection (notably patent and copyright) where the agreements are already in place—although there may still be disagreement over details concerning how the agreements apply to the new technology.

We next consider economic frameworks for analyses of the intellectual property system. We may, for instance, view the system as a social policy that provides for an exchange between society and technological innovators. The innovator provides new technology that benefits society, and society ensures the innovator adequate reward through the intellectual property system. On this view, we are presented with questions about how to set the level of reward. Roughly speaking, the reward must lie between two limits. On the one hand, it must compensate the innovator for the effort of innovation; on the other, it ought not exceed the value to society of the innovation. Much of the modern economic discussion of the patent system is based on a model for the information market developed by Kenneth Arrow in 1962. He suggests that a sufficient level of reward can be reached only with a powerful patent system. Sidney Winter shows how recent and more sophisticated models of the information market suggest that a more limited patent system may be adequate.

Using recent empirical studies, Winter argues that the usefulness of the patent system to innovators (as a means through which to seek reward for innovation) varies with the sort of industry and sort of innovation. Working from studies on the sorts of industries and innovations in which the patent system is most useful, he derives proposals for ways to improve our economic model of the information market and to assess better how the intellectual property system influences research decisions by corporations. We must take into account how the implementation of a patent system itself changes the market conditions in ways that make possible unanticipated new uses of the patent system. For instance, some patents are used to block access to a technology that would permit competitors to step around inventions controlled by prior patents. A sophisticated economic analysis of the interaction between research practice and the intellectual property system must take into account these secondary uses of the patent system.

Patrick Kelly's account of recent changes in patent law is attuned to the interests of practicing researchers in the intellectual property system, one important function of which for researchers is that it provides a form of professional recognition for individual research contributions. Scientists and engineers expect their colleagues to acknowledge their work. Holding a patent is a professionally significant form of accomplishment. Patents are treated as professional credentials, along with such other intellectual accomplishments as degrees, publications, and awards; they are considered seriously in questions of academic promotion and hiring. One way to fight disputes over scientific priority is through legal battles over patent priorities. Apart from any consideration of financial reward, patents hold a very special significance for scientists. Kelly argues as an advocate for scientists concerned with recognition, and he deals with details within the patent system that have interfered with this serious professional concern.

The use of the patent system as a means for recognition is built into patent law. A patent claim must be in the name of the person or persons who actually performed the research and developed the patented item. That person must personally sign the patent application. An inventor may not, for instance, take out the patent in the name of a relative or friend. And even though the patent rights are frequently assigned to a corporate sponsor, the corporation may not make out the claim under its own name. The patent is jointly held by the individuals responsible for the innovative aspects of the claimed invention under their own names. (There are some rare exceptions to this rule, when for instance a sponsored inventor runs off and disappears.)

A patent claim may be successfully challenged if it can be shown that the wrong name or names are on the application. It may be challenged because someone who made a significant contribution to the invention is not recognized as a co-claimant. It may be challenged if someone is listed as a co-claimant who did not make a significant innovative contribution. Thus the patent system is in its very conception not only a means to financial reward, but also a means to professional recognition for innovation.

At the same time that patents are a form of recognition, they are also the basis for rights to profit financially from an invention. Those property rights are more often than not held by the corporate entity that sponsored the research. Under these conditions, we can expect some conflicts to occur between a corporation's need to protect its financial interests and a researcher's expectation that he or she will be recognized for contributions to the invention. Here we have, then, an example of direct conflict between scientific traditions and proprietary rights. It is a conflict with important consequences for whether research practices can remain true to both scientific ideals and corporate needs. Working within the corporate setting, Kelly analyzes how the conflicts are created by conditions built into the intellectual property system, and he makes recommendations for how some of these problems can be alleviated.

Finally, we include in this section a philosophical approach to understanding intellectual property. Although there has been much philosophical discussion of property in general and of issues in social justice that bear on intellectual property, there is little discussion by professional philosophers of intellectual property per se. The situation is changing. Professional philosophers have started to look into intellectual property, inspired by the work of the project that has led to the present volume. In his overview of the philosophical tradition, Alan Goldman draws relevant themes from the broader philosophical literature on property and social justice. His approach is to show how the distinctions found in that literature are reflected in the other studies in this volume and how they can be a basis for further studies of intellectual property.

Most philosophical ethics is roughly categorized as consequentialist or rule based. (The latter is called *deontological* by philosophers.) The consequentialist approach justifies acts, policies, states of affairs, and so on, in terms of the good (however defined) that can be expected to arise as a consequence of acting, or implementing a policy, or establishing a state of affairs. With such an approach, we might view the intellectual property system as a policy that has as its consequence the progress of science (a presumed good). Rule-based ethics evaluates acts, policies, or states of affairs in terms of moral duties and obligations to which we are committed either as human beings, or as professional scientists, or through promises and contracts. For instance, to accept a contract within the intellectual property system may commit us to the system whether that contract or that system is the best arrangement for either ourselves or society. The tradition of philosophical ethics teaches us that these two approaches each have general strengths and weaknesses.

Viewing justifications for intellectual property as typifying one approach or the other, Goldman draws from the broader philosophical tradition familiar criticisms that clarify the strengths and weaknesses of each line of argument. He develops his analysis through a study of progressively more complicated property claims: His simplest case concerns a single innovator working alone without outside funding; he carries the analysis through several complications and concludes with a look at innovations made by a research team attached to a public university and supported with both public and private resources. In each case, he contrasts the lessons to be learned for intellectual property policy under alternative justifications using either a consequentialist or a rule-based argument.

1. General Overview of the Intellectual Property System

Rochelle Cooper Dreyfuss

We are a people increasingly dependent upon innovation. Its fruits support our standard of living and its profits contribute substantially to our prosperity. The central role that invention plays in our economy demands close examination of those factors which influence its creation, distribution, and use. This collection of essays looks at one of these mechanisms: the laws of intellectual property that enable individuals to appropriate new knowledge for their own private benefit. The question of rights to information can be explored from a variety of viewpoints, and while the bulk of this volume takes a largely philosophical and ethical perspective, this paper has the more immediate aim of setting out some of the controversies that are currently working their way through courts and legislatures. By developing a taxonomy of the issues occupying the attention of the legal profession, I hope to provide the reader with a framework for understanding the intellectual property system. The framework will help expose the real-world stakes that are involved in the theoretical inquiries found in these pages.

I have put aside technical questions dealing with the application of the intellectual property statutes to particular cases,[1] and procedural matters common to the administration of the legal system as a whole,[2] in order to concentrate on those problems which best demonstrate how the philosophical issues discussed here come into play in the legal arena. The remaining legal problems are then divided into three categories: 1) questions that require for their resolution an understanding of (if not an agreement upon) the justifications for creating such exclusive rights as patents and copyrights; 2) issues that involve balancing the need for fashioning exclusive rights with other important social concerns; and 3) problems that require explicit investigation into alternative legal strategies for furthering specific goals.[3] Although these categories are not watertight, this breakdown will be useful in delineating the consequences of particular arguments.

The Question of Justifications

One feature of intellectual property that makes it different from some kinds of tangible property is that it can usually be shared. That is, once an invention is perfected, or a manuscript completed, the creator's ability to use it for its intended purpose—to read a research report, or to employ a machine—is usually not diminished by the use made of that same invention by others. Because this distinction between tangible and intangible property is fundamental to the structure of intellectual property law, it is important that its ramifications be fully explored. To do so, compare my ownership of a watch (tangible property) with my invention of a new method for brewing coffee (intellectual property) and ask the question whether each can comfortably be shared.

As to the watch, the answer is usually no. I purchased it in order to ensure my awareness of the time, and this function would be severely impaired if I sometimes had to give the watch over to another. But with a method for brewing coffee, the answer could be yes. Suppose that I invented it to decrease my caffeine intake without significantly sacrificing the flavor of my coffee. My enjoyment of the benefits it produces is in no way reduced when others also use this technology. So long as I can use the technique myself, I drink good coffee and I am spared caffeine. Indeed, society would be better off if everyone knew of my invention: coffee drinkers would be healthier and health care costs would be diminished.

Exclusive rights in intellectual property are, in short, similar in many ways to monopolies in other goods. They have the potential to raise prices, lower output, and produce deadweight social loss as those who could profitably use the innovation (were it priced competitively) forgo purchasing it at the monopolist's price. One could therefore imagine a system of laws that required me to share my invention with the world. Or, less drastically, one could posit a legal regime that held that I may attempt to keep my technique a secret, but if someone manages to learn my method, even against my will, I am powerless to prevent that party from using it. This rule would, in effect, say that if someone reverse engineered the technique (that is, deduced it from an embodiment of the invention, for example, by borrowing my coffeepot and taking it apart) or uncovered it in some more intrusive manner (such as bribing my husband or installing a hidden camera in my kitchen), the public benefit of increased utilization of the discovery should be thought to outweigh my desire to decide for myself with whom I wished to share it. In other words, the legal system should not be structured to enable me to keep my invention as my own exclusive property.

Notice that it would not be so easy to make the same argument with regard to the watch. Sharing my watch will make me late for appointments and class, which will annoy my friends, my business associates, and my students. Since the increased utility to others in having use of the watch is not likely to outweigh my lost utility in forgoing use, there is little public benefit associated with "de-exclusifying" the watch. Thus, there is no point to a legal rule that permits an improper taking—a misappropriation—of it. To the contrary, I should be allowed to count on the legal system to help me keep my watch as my exclusive possession, and in fact the law provides two such remedies: a civil (tort) action to allow me to retrieve my watch, and criminalization of certain interferences with my enjoyment of it.[4]

Since the effect of exclusive possession of intellectual property (*exclusivity*) appears to be a social disutility, the imposition of a private-rights system must be justified, and the justification must be based on some ground other than the practical necessity that supports the laws of tangible property. A variety of rationalizations have been put forward (Machlup 1958), several of which are discussed in this volume. The *natural rights theory* (Davis this volume), for example, holds that the creator has a moral right to capture the benefits of whatever he has invented. Various arguments appear in the literature to support this theory. It is often argued that the right attaches to creators because they have found a previously unused resource, or because they have added their labor to the stock of knowledge. To reap the benefits thus produced, the creator must be able to charge others for the use of his innovation, and this he cannot do unless he has the legal right to exclude those who do not pay.

The *profit-incentive argument* (Goldman this volume) similarly turns on the need to provide a mechanism for financial remuneration. This theory, however, relies on a utilitarian argument. While an inventor's anticipated private enjoyment of his innovation might encourage him to produce, further incentives are thought necessary to provide society with the optimal amount of innovative activity. Market mechanisms will generate an appropriate reward if the innovator is given the right to exclude others, at least for a time. Although he could sell products incorporating his innovation without this right, he would compete at a disadvantage. Since those who copy the invention will not themselves have spent time and money developing it, they will be able to offer the invention to the public at a lower price. Although the public would benefit from the price reduction offered by these free riders, the inventor will make fewer (if any) sales, he will not be able to recoup his costs, and he will lack the financial incentive to continue to innovate.

Other justifications stem from noneconomic premises. The *exchange-for-secrecy rationale* (Kuflik this volume) states that without a legal right to prevent others from copying his invention, the creator may be tempted to keep it a secret. If he does, others may unknowingly duplicate the effort that went into creating it. If, in contrast, the inventor can rely on legal rules to prevent those who learn the secret from themselves exploiting it, or to seek compensation from those who do learn and use it, then the inventor will be willing to disseminate the invention more widely and disclose its details to others. This disclosure will, in turn, enable others to use the ideas the invention embodies to extend the frontiers of knowledge. Even though the public will pay more for a particular invention because the creator owns an exclusive right, the extra cost is less of a burden than not having the secret revealed at all.[5]

The *quality-control principle* (Weiner this volume) looks at the exclusive right not so much as an element in encouraging invention or disclosure, but more as a method for protecting the innovation once it is released. By giving the holder power to control how the invention is used, exclusive rights enable him to maintain its integrity. He can, for example, use this right to prevent others from distorting or mutilating his work, thereby diluting its quality.

The *prospecting theory* shares some of these quality-promotion elements. It argues that one value in a system of exclusive rights is that it concentrates research. Like a miner who owns his mining claim, the holder of an exclusive right has the incentive to develop fully his ideas. And since anyone else who wants to pursue work in that field necessarily approaches him first, the holder comes to possess comprehensive knowledge of how the field is unfolding and can help maintain an "orderly market" in its further development (Kitch 1977).

At the heart of many pivotal disputes in intellectual property law are questions as to whether exclusivity is justified in a particular case, and if so, on what theory. How these questions are resolved determines whether a particular endeavor will receive legal protection at all[6] and whether it will receive that protection under the patent system,[7] the copyright laws,[8] or some other state or federal regime.[9] In the end, the choice of justification shapes both the rights of creative individuals and the quality (and quantity) of the product that passes to the public. (Samuelson, Buttel and Belsky this volume).

The centrality of the justification issue to the structure of intellectual property law can be demonstrated by looking at how the legal system handles novel claims to protection. The emerging right of publicity provides one such illustration. Until recently, it was quite clear that anyone was free to tell the story of another person's life. Legal protection, to the extent that it was available at all, protected only the right to privacy. Thus, if the plaintiff's interest was in withholding intimate details about his life, he could (at least sometimes) invoke the protection of the court.[10] However, once information about a person became publicly accessible, it was permissible for anyone to use it for any purpose, including commercial benefit. Thus, so long as information was gleaned from purely public sources, unauthorized biographies as well as sociological and historical studies, were not actionable; similarly, plays, movies, and television shows could be made of the lives of interesting persons without their permission.

Within the last few years, however, a new strand of analysis has appeared. The claim has now been made that there is an interest apart from privacy that is infringed by these works: that even the public aspects of a person's story are his property, and that the law should protect his exclusive right to utilize that property.[11] Although no full-blown attempt to claim exclusivity in a life story has yet to reach the courts,[12] Estate of Presley v. Russen,[13] which concerned the right to impersonate the late Elvis Presley, provides a vehicle for considering how such a claim will be evaluated.[14]

Clearly, there is a public interest in facilitating impersonations. They enable older generations to relive earlier experiences and share their culture with younger audiences. Impersonation is, in addition, a fascinating art form of its own. Should, then, the public be given free access to Presley's act? The answer depends on which rationale is used to justify the right to exclusivity, for the theories point in contrary directions. If, for example, exclusive rights are thought necessary to promote disclosure of secrets, then the Presley estate should lose. Presley is, after all, dead, and has no further secrets—no new songs or performing ideas—to reveal. The profit-incentive rationale leads to the same conclusion. It could be argued that as long as the impersonator makes clear that he is not providing the audience with the genuine Elvis Presley, the money that he makes cannot be counted as part of the incentive that motivated Presley to produce.[15]

At the same time, however, the estate wins on any of the other theories. Since the impersonator repeats activity that originated with Presley, the estate, as the representative of his interests, could make a substantial moral claim to any natural right Presley had to his act. Furthermore, the impersonations have an impact on the public perception of Presley. The argument that it is desirable to appoint a caretaker to regulate the market for the creator's product and protect its integrity would seem to apply with full force to impersonation.

The choice of justifications affects not only whether an exclusive right is created, as in the above illustration, but its contours as well. For example, if the right of publicity is considered property, then it may seem obvious that it should be treated for inheritance purposes just like any other property.[16] Thus, the owner of the right should be able to dispose of it by will, and if he fails to do so, it should pass to his heirs under the laws of intestacy.

However, close inspection of the justifications supporting exclusivity reveals that another solution may make more sense. Perhaps the right to exclusivity should be recognized during the creator's lifetime, but it should be extinguished upon his death. As applied to the Presley case, the argument under the profit-incentive rationale is as follows: While he is alive, the entertainer and the copyist are in competition with one another. Although the impersonator may not offer precisely the same experience as Presley, he does offer a similar product at a lower price. If a part of the audience is willing to accept the substitution, Presley's profit is diminished. To establish appropriate monetary incentives, the right of publicity must be recognized. After death, however, the creator loses the ability to create, so an incentive is not needed, and the right could be terminated.[17]

Because of its structural implications, the choice of justification also plays a crucial role in many cases involving long recognized exclusive rights, such as federal copyrights and patents. Although these rights are based on article 1, section 8, clause 8 of the Constitution, which directs Congress "to promote the progress of science and useful arts, by securing for limited times to authors and inventors the exclusive right to their respective writings and discoveries," the Constitution does not adopt any particular rationale for believing that progress will be promoted through exclusivity. Thus, in a sense, the issue of justification is somewhat open even in the case of federal rights. It is, however, probably fair to say that the Supreme Court has read the profit-incentive theory into the law,[18] and this view has, in turn, shaped the extent to which creators can avail themselves of federal law to oversee the quality of their product or further their prospecting interests in continuing to regulate exploitation of their innovations.

The Copyright Act, for example, creates a series of restrictions on the author's right to control, including the first-sale doctrine (which allows the copyright holder to control only the first sale of any copy of his work),[19] the principle of fair use (which tolerates certain unauthorized, uncompensated uses of a copyrighted work),[20] and compulsory licenses (which permit certain unauthorized uses at nonmarket prices).[21] These limitations stem in part from legislative judgments that financial considerations are of paramount concern, and that the number of profit-making opportunities created by the law should be no greater than those which will stimulate the optimum amount of innovative activity. Apart from doubts about whether Congress was sufficiently generous to innovators,[22] these doctrines can also be criticized for ignoring the rationales that are directed at protecting the creative environment, and for depriving creators of control over the integrity, impact, and public perception of their work.[23] Indeed, even the limited duration of copyright is problematic in this regard, as the debate over colorization of old motion pictures demonstrates.[24]

Some states do not disregard the other justifications for exclusive rights. Thus, the unfettered right of a buyer to dispose of an original "copy" of a work of fine art has been modified in several states by statutes that enable artists to protect their works from mutilation and destruction.[25] Similarly, state contract, tort, and trade secrecy laws do, in some circumstances, further the objectives underlying the nonpecuniary rationales.[26] Surprisingly, even state unfair competition laws, which are largely designed to protect the efficient conduct of the marketplace, sometimes fill the interstices created by the financial emphasis of federal law. By permitting the creator to prevent others from making changes in his product, these laws allow him to maintain the quality of the material that becomes part of the public domain.

But when there are state laws that establish wider rights, innovators must deal with another, more subtle, question concerning justifications: does the adoption of one theory of protection exclude the application of the other rationales? The exchange-for-secrecy rationale is, for example, somewhat inconsistent with a regime of trade secrecy that helps inventors keep their discoveries to themselves. Accordingly, it is not surprising that for a time the Supreme Court thought the national interest required preemption—invalidation—of state laws providing exclusivity to innovations that are not protected under federal law. In Sears, Roebuck & Co. v. Stiffel Co., for example, the Court held that "to allow a State by use of its law of unfair competition to prevent the copying of an article . . . would be to permit the State to block off from the public something which federal law has said belongs to the public."[27]

But since the high-water mark of preemption in the 1960s, the Court has substantially retreated from this position and has, in more recent years, sometimes allowed states to offer protection in circumstances not covered by federal law.[28] In part, the trend toward greater flexibility has itself stemmed from the perceived need to provide a profit to motivate research. In Kewanee Oil Co. v. Bicron Corp.,[29] for instance, the Court permitted a state to protect through its trade secrecy law an invention that could have been, but was not, patented. Although the Court recognized that enforcing the trade secrecy law conflicted with an exchange-for-secrets rationale, it reasoned that a state may legitimately decide to use the profit motive to encourage innovation and offer a mechanism of exclusivity to augment the patent law.

Recent cases, however, hint that even nonpecuniary rationales may yet capture the Court's imagination. For instance, the Court accepted a quality-control justification for allowing President Ford to utilize copyright law to determine the time when his memoirs became publicly available.[30] Although the Court recognized that its decision would impinge on the public's access to a work of great national interest, it reasoned that if an author were unable to block the publication of his unfinished works, there would be nothing to prevent the dissemination of inaccurate, unpolished material, and this, too, would work against the public weal. If this trend toward recognizing multiple justifications for exclusivity continues to flourish, the artists' rights provisions, which have not yet been tested in the Supreme Court, may survive preemption challenges. Moreover, nonpecuniary justifications could eventually draw the attention of Congress, paving the way to federal protection expressly premised on concerns for the creative environment.[31]

The Question of Balancing

Debates over the application of doctrines that limit the creator's ability to protect his work stem from more than a dispute over the appropriate justification for exclusivity. Even if a particular objective were universally accepted, questions would remain concerning the extent to which the agreed goal should be compromised in favor of other social interests.

The limitations in copyright law mentioned above are examples of provisions meant to balance the creator's needs against the public's interest in access to his work. The Copyright Act recognizes, for example, that there may be circumstances in which society benefits by use of a work, but that it would be difficult (if not impossible) to secure the author's permission to put his work to that use (Gordon 1982). To ensure public access for these important purposes, the act encroaches on the creator's interests and allows others to utilize the work without his authorization. In these circumstances, the author's remuneration is either determined by nonmarket forces, as in the case of compulsory licenses, or relinquished entirely.

The fair use doctrine is an example of the latter solution. Its application was at issue in Maxtone-Graham v. Burtchaell,[32] a case involving the right of an opponent of abortion to use excerpts from interviews given, at high personal cost, by women who had undergone abortions and wished to preserve the right of others to make similar choices in the future. Even on a a theory of quality control (or exchange for secrecy), the Second Circuit was correct in holding that the quotations were a fair use of these works. Although the decision allowed Burtchaell to use the interviews to support a cause that the interviewer and interviewees opposed, the right to quote is essential if the public is to discuss intelligently the contributions that the creator made and to build upon them to further extend the frontiers of knowledge.[33] Permitting Maxtone-Graham to withhold permission to use these interviews would, in effect, allow her to prescribe the terms of the debate on an important social issue.

Because the national agenda is always in flux, the strategy for mediating between private and public concerns must be reexamined whenever new interests emerge or old interests acquire new importance. Consider, for instance, the question of protecting ideas and facts, an issue that has become prominent as technology and information services have become increasingly significant components of the gross national product. Because free use of facts and ideas has always been considered a public necessity, the law limits patents to applications of ideas and copyrights to original expression of ideas or facts. In that way, the ideas that underlie a patented innovation are available to other inventors, and the ideas and facts expressed in a copyrighted work can be utilized by other authors. If, however, the information and technology industries are to continue to thrive, it may be necessary to rethink this traditional regime.

The tension between encouraging the development of new ideas, or works with high factual content, and safeguarding the public's right to discuss freely ideas and facts is nicely demonstrated in the conflicting opinions of the Second Circuit and the Supreme Court in Harper & Row Publishers, Inc. v. Nation Enterprises,[34] the case involving the *Nation*'s unauthorized publication of a portion of President Ford's memoirs before its official publication date. The Second Circuit reasoned that if the copyright on these memoirs were to prevent discussion of such issues as Richard Nixon's pardon, there would be little public benefit in encouraging President Ford to write. Accordingly, it grafted a public-figure exception onto the copyright law and held the work largely uncopyrightable. The Supreme Court, on the other hand, recognized that if the Ford memoirs were freely available, then future presidents would not have the financial incentive to set down their recollections. It therefore reversed the Second Circuit and refused to find Ford's status a bar to the copyrightability of his work.

The arguments heard in the Ford case are also made in regard to the protectability of those works which form the backbone of the information industry, such as computerized data bases. The prevailing approach in these cases has been to decide whether a work is copyrightable by looking at the degree of creativity it embodies. As a result, the more closely a work reflects actual facts, the less likely it is to be protected.[35] This focus on inventiveness neatly protects public access to the factual content of the material, but it largely ignores the real issue of how to stimulate the effort of collecting. Without the labor, the public would not, after all, have the facts so conveniently arrayed and available for use.

State law has, once again, sometimes stepped into the breach by finding that the sale of another's work without the author's permission is an unfair method of competing even when the work is not copyrightable. For example, the New York State Court of Appeals found, as a matter of state law, that the sale of opera music recorded from a radio telecast of the Metropolitan Opera was a misappropriation of material that belonged to the Met. Although sound recordings were not at the time of the case covered by copyright law, the court thought it necessary to protect the Met against activity that threatened its capacity to stage further productions.[36]

These state-based rights are, however, of fairly limited significance. They are available only in states that recognize them. Furthermore, they are, as has been shown, vulnerable to preemption claims, and sometimes tend to turn on distinctive features that may not be found in the entire class of cases raising similar concerns. If development of the information industry is considered vital, systematic federal legislation may be needed, even if that protection requires some intrusion into the public access interest.[37]

Similar arguments apply to patent law. As with the fact/expression dichotomy of copyrights, patent law's distinction between protectable applications and nonprotectable ideas is easier to describe than implement. Excessive sympathy with the public-access aspect of the problem may, in the end, provide insufficient stimulation to basic research, especially in fields like computer programming, where the applications of the work are not easily separated from the ideas contained.[38]

The notion that changes in the national agenda require readjusting the balance between proprietary and public interests is also evident in the controversy over the "Nine No-Nos" of patent licensing. For some time a fixture of the Justice Department's antitrust policy, the No-Nos were a list of licensing practices (such as resale price maintenance, tie-outs[39] and tie- ins[40]) that were, in the government's view, anticompetitive (Lipsky 1981). Although the department explained the restrictions imposed by the No-Nos as necessary to prevent the patentee from extending his patent grant beyond its terms—to restrain the patentee from capturing more than the benefit conferred by his invention—this rigid approach has recently come into disfavor (Rule 1986).

Theorists now argue that the No-Nos went far beyond the quantum of regulation required by antitrust policy. These thinkers claim that it is not likely that anyone (including the licensee or his customers) would pay more for the invention than it is worth, no matter how the patentee structured the transaction. Accordingly, except in rare circumstances, it is not possible for the patentee to extract from the public more than the value that the invention created. Under this analysis, the effect of the No-Nos is to shortchange the patentee and prevent him from fully exploiting his patent rights. Although the invention may then be cheaper to its consumers, the No-Nos reduce the financial incentive available to stimulate innovative activity. Since the department now considers innovation an important national goal, and because it believes that innovation is best fostered by allowing inventors to exploit their works fully, it has greatly relaxed and almost abolished the No-Nos (Andewelt 1985).[41]

Whether this is the final word on the licensing issue remains to be seen. As patent rights become more valuable, patentees may be able to use their dominant and unique positions to discourage challengers and suppress newer inventions. If the result is a net loss in progressiveness, the balance between patent rights and antitrust limitations may have to be examined once again.

The rights of innovators must also be weighed against newly emerging public interests. Compatibility is one such concern, for it has become increasingly evident that access alone is not always enough to put innovations to their best use (Davidson, Samuelson, Lemin this volume). The distinction between access, on the one hand, and access plus compatibility, on the other, is illustrated by my search for a personal computer for my home. In theory, any computer on the market should be of interest to me. In fact, however, I will look at only one subclass, and that is comprised of computers that are compatible with the computer in my office at New York University (NYU). Only if I buy a computer for home that runs the word-processing program that I use at school will I be able to work at home on files I create in my office. The same concern arises when I purchase software, printers, and other peripheral equipment. And the interest is not mine alone, for every manufacturer wishing to compete for my computer, software, and peripherals business is concerned with making products compatible with the computer that NYU purchased for my office.

Patent and copyright laws have traditionally safeguarded the interest in compatible products through their deposit and specification requirements. The Copyright Act requires the copyright holder to deposit copies of his work in order to obtain the right to sue for infringement;[42] the Patent Act requires the patentee to reveal in a publicly available specification all the information necessary to duplicate his invention.[43] During the term of protection, these provisions guarantee that sufficient information is revealed so that others can manufacture products compatible with the innovator's invention. After the term of protection has expired, they ensure that competitors know enough to compete effectively.

The current enthusiasm for stimulating innovation has, however, led to a relaxation of these rules for new technologies. The Semiconductor Chip Protection Act of 1984,[44] for example, allows the inventor of an integrated circuit to obtain exclusive rights to his design without entirely revealing the elements that make the design operative.[45] Similarly, the Copyright Office now permits software producers to withhold selected portions of their programs from their deposit copies.[46] Although these rules could be challenged on the ground that they undervalue compatibility concerns, analogous claims against trade secret holders have so far met with little success.[47] To date, the accepted view is that disclosure is not necessary to maintain vigorous competition, but new technologies and a greater appreciation of the benefits of compatibility may lead to different results in the future.

The environment is another nascent focus of public interest. In recent years, a series of right-to-know laws have been enacted at both state and federal levels so that the impact of innovative activity on health and safety can be monitored. The federal Occupational Safety and Health Administration demands, for example, that chemical manufacturers label containers of hazardous chemicals with the chemical's identity. Employees, their representatives, and health professionals have a right to further vital—and secret—data about these chemicals in specific circumstances.[48] Some states have gone even further. New Jersey requires employers to prepare environmental surveys indicating the hazardous chemicals that are found in the workplace. These surveys are submitted to various governmental authorities, and are available in part to employees.[49] Although these statutes offer partial protection for trade secrets (by, for instance, requiring persons receiving information to sign confidentiality agreements), it is clear that secrecy is severely compromised by any regulation that requires circulation of trade secrets to parties who are beyond the control of the inventor.

Disasters such as Union Carbide's in Bhopal, India, have focused the public's attention on the need for more information about activities that occur in its midst. But despite the appealing nature of the claims that have led to this legislation, "de-exclusifying" protectable intellectual property may have the unintended consequence of diminishing the opportunities for recouping the costs of invention. These enactments are too recent to have been extensively tested in the courts,[50] and insufficient time has elapsed in which to observe their impact on inventive behavior. It will be interesting to see where the nation ultimately strikes the balance between the new social policies furthered by these laws and the accelerating concern for stimulating creativity.

Alternative Schemes

It has been pointed out that the current regime of intellectual property laws is not immutable, and that alternative schemes for protecting innovators could be adopted (Kuflik this volume). The problems noted thus far perhaps hint that the legal system should be actively engaged in the enterprise of searching for better ways to promote innovation.

There are two directions from which to approach the problem. One possibility is to tinker with the current system (Kelly this volume). That is, once a particular difficulty is detected, the legislature could simply alter the statute to eliminate the problem.[51] This solution has, of course, been followed to a large extent. For instance, three years ago Congress extended the patent term for drugs subject to delays due to regulatory activities,[52] and has considered similar action for agrochemical products and animal drugs.[53] A provision that would bring copyright law into closer alignment with the European quality-control vision of authors' rights is under active contemplation,[54] and the Reagan administration advanced initiatives to clarify the rights of patent licensors.[55] Future problems could (and, indeed, will) be solved in the same manner. For example, should a patentee develop a socially important invention (such as a cure for cancer—or a better technique for brewing coffee) that she refuses to exploit, Congress would undoubtedly quickly import the compulsory license strategy of copyright law into the patent system and enact a statute permitting others to use the invention upon payment of a fee set by the government.

A more drastic approach to the problems surrounding the current legal regime would be to abandon the old system and start afresh. The United States could, for example, move to a patronage system whereby innovators were subsidized by government, social organizations, or other wealthy entities. James Madison at one time suggested that the government determine the social value of every new invention and award an amount proportional to that value to the inventor. A bounty system has also been proposed that would allow the government to purchase innovations and dedicate them to the public. If the government chose not to buy, then the inventor would be given protection from free riders through a system akin to the present one (Jewkes, Sawers, and Stillerman 1969; Scherer 1980; National Academy of Sciences 1962).

To some extent, these ideas are currently being implemented. Government grants and tax write-offs are forms of patronage, and government purchase orders are a variation on the bounty theme. Relying exclusively on these alternatives would, however, introduce a series of new problems. Unlike the market mechanism at the heart of the current system of intellectual property protection, these schemes require someone to make ex ante evaluations that can be difficult and are sometimes dangerous. Patronage systems, for instance, single out certain inventors to receive financial backing, and so require someone to prejudge the abilities of the applicant pool. As experiences with government funding have demonstrated, maverick thinkers who are not affiliated with well-accepted organizations tend to be ignored in these processes. In countries where subsidies are funneled only to those who adhere to particular political philosophies, the results for science have been alarming.[56] Although bounty systems are somewhat less vulnerable to politicization, they also require evaluations of the benefits of the invention. An inventor who takes an unconventional approach is still unlikely to enjoy a reward under such a system.[57]

Congress has occasionally tried to capture the benefits of both tinkering with the old and starting anew. The Semiconductor Chip Protection Act represents one such attempt. A special law enacted to meet the needs of a single technology, the Chip Act is a fascinating experiment in intellectual property legislation. In some respects, such as its use of a market mechanism to reward invention, the act closely resembles copyright and patent law. In other ways, however, Congress has written on a clean slate. Concepts such as fair use have largely been scrapped; instead, the act secures the public's right of access by allowing others to examine and copy protected chips for the purpose of pushing the technology forward.[58] At the same time, the act safeguards proprietary rights by relaxing the requirement of patent law that the application fully reveal all the secrets in the invention. Other special features of the Chip Act—such as a change in the term of protection, protection for those who violate exclusive rights innocently, and a new originality standard—also represent attempts to develop novel solutions that are highly responsive to the needs of the semiconductor industry.

But while this specialized scheme may be a tremendous advantage to integrated-circuit producers, developments of this type present many new challenges to the legal system. The primary one is determining which technologies merit this special attention. It has been forcefully suggested that the computer industry should also receive tailored protection because it suffers many of the same problems that led Congress to enact the Chip Act (Samuelson this volume). But other fields have similar difficulties with existing law. Industrial designers, for instance, are also hampered by many of the concerns that motivated the Chip Act.[59] Other fields, such as microbiology and superconductivity, present new problems that cry out for unique solutions (Lemin, Davidson, Kelly,

Samuelson, Winter, Buttel and Belsky this volume). By providing a new option for Congress to consider, the Chip Act may have done no more than introduce yet another kind of tension into the system.

And if the law continues to move toward more specialized measures, other problems will come into focus. Fine-tuned legislation requires a wait-and-see period during which the needs of the new technology are given time to crystallize. In the interim, the prospect of—and the contours of—the new legislation may be unclear, and innovations that occur before the new scheme is announced may become even more vulnerable to copyists. The advantage given free riders during the time before new legislation is enacted may, in fact, lead to the undoing of the first innovators. Because of these uncertainties, inventors and their financial backers may prefer in the future to concentrate their efforts on those technologies which are subject to laws that are better defined. As a result of Congress's increased willingness to think about novel legislation, research could, paradoxically, end up moving away from those fields where the claim for specialized legislation is especially strong.[60]

Even more serious are the international implications of these schemes. As the world becomes smaller—or as inventors depend more on world markets to recoup their costs—international protection becomes increasingly important. In recognition of the need to assure its citizens protection in important markets, the United States has entered various international, regional, and bilateral treaties.[61] But while these treaties facilitate acquisition of patents and copyrights in signatory countries, they do not apply to alternative protection schemes. Thus, every time the United States (or a foreign country) heeds the cries of a new industry and enacts specialized legislation, new international agreements have to be forged. The experience of the Chip Act is that international accommodation is a slow process. Even after the act was several years old, not a single foreign country had adopted a measure that the United States considered as protective as its own.[62]

In the final analysis, it may turn out that the benefits of starting afresh, considerable though they may be, are radically outweighed by the short-term burdens endured during the period when protection is uncertain, and by the costs incurred in the international arena. The law is a conservative enterprise, and this is especially true when the issue it addresses is the creation of interests that are intended to motivate future performance.

Empirical Research

A puzzling feature of all of this is that many of the controversies described here could be better understood (if not resolved) by empirical research. Did Elvis Presley count on being able to exploit his persona after his death? Are ex-presidents—or scientists—motivated by financial considerations? Do people rely on trade secrecy when patent protection is available? Would products be more convenient if compatibility concerns were more vigorously protected? Several of the essays in this volume lament the lack of empirical research, and I conclude this essay with a few brief thoughts as to why such data, at least in the form of surveys, are not forthcoming.[63]

Probably the most significant problem is that these studies are difficult to conduct in a manner that lends itself to unambiguous interpretation. As it turns out, the answers obtained on surveys depend very much on who responds to the questions. Is the patent system critical to research and development? It is not surprising that large firms usually claim it is not. Since these firms are vertically integrated—they control research and development as well as production and distribution—they have at their disposal a variety of tactics for exploiting their inventions, many of which do not rely on patent rights. Small firms, on the other hand, lack the capacity to produce or distribute their inventions themselves. To earn financial return on their discoveries, they must rely on licensing. Although patent rights are not required to obtain royalties for the use of an invention, the amount of royalties will depend on the benefit that the licensee expects to obtain. Without patent protection, small firms cannot reveal enough information to potential licensees to negotiate for the best price for the use of their inventions. Thus, small, nonintegrated firms would tend to report that patent rights are significant to their research decisions (Dreyfuss 1986).

Surveys will be imperfect even if they are corrected for structural problems. Questions that require people to predict how they will behave if the legal regime were different demand imagination and a sophisticated understanding of both the current and the proposed legal rules. Many of the potential targets of surveys lack a firm grip on the details of the regime under which they are now operating; they are unlikely to spend a great deal of time learning about theoretical possibilities. Some authors have little understanding of what it is that motivates their publisher to buy, rather than steal, their work. Many nonlawyers can barely credit the notion that the patent (or copyright) system could be abolished. Parties who have had experience with the current system (such as lawyers, litigants, or witnesses) have a different perspective from those who have not seen the system in operation. Research and development managers have a different view from the patent lawyers, the bench scientists, or the advertising executives employed by their own firms.

Finally, this is an area where, by definition, secrecy is important. Surveys attempting to measure the extent to which trade secrecy is relied upon are notoriously unsuccessful because most firms regard *whether* they rely on trade secrecy as itself a secret.[64] Similarly, firms are extremely reluctant to reveal the extent to which they believe their rights are being infringed. They are afraid that if the numbers appear too high, others may believe that the validity of the underlying claims is in doubt. The direction of future research, how the fruits of creativity will be exploited, the problems anticipated with existing law, are similarly all sensitive items of information that firms hesitate to disclose.

The national agenda will continue to shift as new discoveries are made and public policies emerge and recede in importance. Perhaps the future will also bring better methods for dealing with factual uncertainty. In the meantime, the structure of intellectual property law and the balance that it strikes between public and private rights will depend upon the proficiency with which we wield the analytical tools currently at our disposal. By bringing together the work of lawyers, economists, scientists, and philosophers, this volume makes impressive strides in illuminating those issues which are especially worthy of consideration.

Notes

1. See, for example, Hodosh v. Block Drug Co., Inc., 786 F.2d 1136 (Fed. Cir.) (considering whether an ancient Chinese folk medicine text is prior art within the meaning of the Patent Act, 35 U.S.C. sec. 102[b], *cert. denied* 107 S.Ct. 106 [1986]; Past Pluto Productions Corp. v. Dana, 627 F. Supp. 1435 (S.D. N.Y. 1986) (deciding that Statue of Liberty hats lack sufficient originality to qualify for copyright protection as a derivative work under the Copyright Act, 17 U.S.C. sec. 103).

2. See, for example, Dennison Mfg. Co. v. Panduit Corp., 475 U.S. 809 (1986) (discussing the scope of appellate court review of trial court patentability decisions); Young v. U.S. ex rel Vuitton et Fils S. A. 107 S.Ct. 2124 (1987) (holding improper the appointment of an interested attorney to prosecute contempts of trademark injunctions).

3. Nonetheless, this breakdown is useful for delineating the consequences of particular arguments.

4. This is not to say that there may not be other, nonutilitarian, justifications for private tangible property rights.

5. An example may be helpful here. Say that I wish to exploit my coffee-brewing technique. Because of its popularity, I will have to open two coffee-brewing factories if I am to satisfy demand. But since I can be at only one place at a time, I can run the second factory only if I am willing to tell my secret to a foreman. Whether I open two factories or only one factory will depend on my legal rights to the secret. If I can rely on the law to prevent the foreman from double-crossing me and opening a rival factory, then I will reveal the secret and open two factories. If, on the other hand, I have no legal recourse against the foreman, then the only way to prevent him from going into competition with me will be to withhold the secret. I will, accordingly, open only one factory, leaving substantial unsatisfied demand.

Crucial to understanding this theory is recognizing that no legal rule is required to maintain a *real* secret. Patent law and trade secrecy law come into play only when something is no longer a real secret. Thus, they virtually always operate to prevent people who actually do know the innovation from practicing it without permission (and usually payment).

6. It could be argued that innovations inimical to health and welfare should not be protected so that further developments are discouraged. Such a claim has recently been made with regard to the patenting of multicellular organisms; see Schneider (1987). This issue is discussed in A. J. Lemin's article on microorganisms (this volume) and Diamond v. Chakrabarty, 447 U.S. 303 (1980). See also Cranor's discussion of body parts (this volume).

7. Federal law creates three kinds of patents. The inventor of a "new and useful process, machine, manufacture, or composition of matter" is eligible for a utility patent, which creates rights against those who would make, use, or sell the invention without permission (35 U.S.C. secs. 101, 271). The design patent provides similar protection for the inventor of a "new, original, and ornamental design for an article of manufacture" (sec. 171). A plant patent allows one who discovers and asexually reproduces a "distinct and new variety of plant" to exclude "others from asexually reproducing the plant or selling or using the plant so reproduced" (sec. 161). Patents create exclusivity for only a limited time, generally seventeen years (sec. 154).

8. A copyright protects "original works of authorship fixed in any tangible medium of expression" against unauthorized reproduction, distribution, public performance, or display. The copyright holder is, in addition, given the right to exclude others from preparing derivative works (such as translations) based upon the original work (17 U.S.C. secs. 101, 106). Like patents, the term of copyright protection is limited, generally to the life of the author plus the fifty years following his death (sec. 302[a]).

9. State trade secrecy and unfair competition laws create exclusive rights by providing remedies against those who discover a secret innovation improperly (for example, through bribery or espionage) or compete unfairly (for example, by using the inventor's trademarks to convey incorrectly to others that the product they are selling is that of the inventor). In theory, these rights extend indefinitely; in practice, however, the protection generally dissipates as others independently invent or reverse engineer the secret. Furthermore, unfair competition claims cannot usually be asserted after the trademark holder has ceased doing business.

10. For example, in Froelich v. Adair, 213 Kan. 357, 516 P.2d 993 (1973), a detective was held liable for bribing a hospital orderly to obtain hair combings from a patient's brush, the court reasoning that the plaintiff had a right to prevent others from intruding into his personal affairs.

11. Thus, in his seminal work, Prosser considered the right of publicity to be an aspect of privacy (see Prosser 1971); this analysis was carried over into the *Restatement of Torts* by the American Law Institute (1976, sec. 652A). Various commentators have, however, begun to realize that privacy, which protects information from being revealed, and publicity, which protects the right to exploit information that has been revealed, are quite distinct from one another; see, for example, Kwall (1983). Indeed, even the *Restatement* contains hints that the two rights are only tenuously connected; see American Law Institute (1976, sec. 652A, comment b).

12. There are, however, clear indications that sooner or later the attempt will be made. The Baby M case on surrogate motherhood has, for example, spawned a satellite

controversy over whether Baby M can (through her guardian) prevent the natural mother from telling her story for profit without paying royalties to her; see Hanley (1987). Similarly, Elizabeth Taylor threatened litigation to prevent the American Broadcasting Corporation from making a docudrama of her life (see Lewin 1982, quoting the actress as saying, "I am my own commodity. I am my own industry."). See also Salinger v. Random House, 811 F.2d 90 (2d Cir.) (discussing whether a biographer can, consistent with the copyright laws, use material found in a library to write a convincing and lively account of a reclusive author's life), *cert. denied*, 108 S.Ct. 213 (1987).

13. 513 F. Supp. 1339 (D. N.J. 1981).

14. Numerous other impersonation cases have also been litigated, including claims by Bela Lugosi, the estate of the Marx Brothers, Jacqueline Onassis, Cher, and Woody Allen. Many of these are described in Allen v. National Video, Inc., 610 F.Supp. 612 (S.D. N.Y. 1985). See also Apple Corps Ltd. v. Leber, 229 U.S.P.Q. 1015 (Cal. App. Dep't. Super.Ct. 1986) (discussing *Beatlemania*, as an infringement of the Beatles' publicity rights).

15. It is useful to compare this case with Zacchini v. Scripps-Howard Broadcasting Co., 433 U.S. 562 (1977), where the Supreme Court upheld a right-of-publicity claim. In that case, however, the defendant had filmed the plaintiff's entire act and showed it on television, thus arguably exhausting the demand for the plaintiff's product. In contrast, the demand for the "real thing" is not likely to be satisfied with an impersonator.

16. See, for example, Tennessee ex rel. Elvis Presley Int'l. Memorial Found'n. v. Crowell, 733 S.W. 2d 89 (Tenn. Ct. App. 1987). Accepting a property rather than a privacy rationale for the right of publicity, this court looked to the justifications for creating exclusivity in order to determine whether the right should descend to Presley's heirs.

17. Although it is true that after his death the creator's heirs may find their ability to sell remembrances of him (such as records) impaired by the offerings of the copyists, it is not likely that an innovator would count heavily on this return in deciding whether to innovate. Thus, the marginal motivational value of the return after death may not compensate the public for forgoing the posthumous performances of copyists.

Similar arguments could be made under some of the other theories. The State of Florida, for example, has accepted a rather sophisticated argument. It recognizes the possibility that a creator may be motivated by the desire to provide for his spouse and children. Thus, Florida has enacted a statute providing a right of publicity that endures during the life of the creator, and also during the lifetimes of his spouse and children. When they die, the right is extinguished. See, for example, Southeast Bank v. Lawrence, 66 N.Y. 2d 910, 498 N.Y.S. 2d 775, 489 N.E. 744 (N.Y. Ct. App. 1985).

18. See, for example, Harper & Row, Publishers, Inc. v. Nation Enterprises, 471 U.S. 539, 546 (1985) (stating that "the rights conferred by copyright are designed to assure contributors to the store of knowledge a fair return for their labors"); Twentieth Century Music Corp. v. Aiken, 422 U.S. 151, 156 (1975); and Mazer v. Stein, 347 U.S. 201, 219 (1954) (speaking of the "economic philosophy behind the clause empowering Congress to grant patents and copyrights").

19. 17 U.S.C. sec. 109(a). Since federal law defines *copy* to include the original embodiment of a work (17 U.S.C. sec. 101), the first-sale doctrine arguably permits the

owner of a painting to deface it, the owner of a bronze sculpture to melt it, or the owner of an original manuscript to burn it. It also permits the owner to rent his copy to those who would otherwise purchase it from the copyright owner; cf. Columbia Pictures Industries, Inc. v. Redd Horne, Inc., 749 F.2d 154 (3d Cir. 1984) (enjoining videotape rentals, but only when they can be classified as public performances).

20. 17 U.S.C. sec. 107 permits such use "for purposes such as criticism, comment, news reporting, teaching, scholarship or research." The statute provides a series of factors to decide whether a particular use is fair. These include the purpose of the use (whether it is commercial or not-for-profit), the nature of the work (such as whether it is factual), the substantiality of the use, and the effect of the use on the market for the work.

21. The act specifies which uses are subject to compulsory licenses. These include retransmissions of copyrighted works by cable systems (17 U.S.C. sec. 111), and placement of records in jukeboxes (17 U.S.C. sec. 116). Royalties are set by the Copyright Royalty Tribunal, an administrative body created by the act (see secs. 801–803). Although the tribunal attempts to determine market rates, it is possible that the rates they set are lower than the amount that would have been charged had the owner been free to refuse to negotiate with the user.

22. This was, in part, the issue in Sony Corp. v. Universal City Studios, Inc., 464 U.S. 417 (1984), which presented the question whether motion picture producers have the right to royalties when their work is videotaped from public broadcasts.

23. See, for example, Shostakovich v. Twentieth Century–Fox Film Corp., 196 Misc. 67, 80 N.Y.S. 2d 575 (1948), *aff'd*, 275 A.D. 692, 87 N.Y.S. 2d 430 (1949) (refusing to enjoin the showing of anticommunist film using the works of Dmitri Shostakovich and other Soviet composers despite the claim that association of their music with an anti-Soviet theme injured their reputations). Cf. Henry Holt & Co., Inc. v. Liggett & Myers Tobacco Co., 23 F. Supp. 302 (E.D. Pa. 1938) (scientific paper quoted by cigarette company in a way that implied the author was hired to find cigarettes safe).

24. See Bennetts (1986) describing Frank Capra's effort to prevent computer colorization of his classic black-and-white film, *It's a Wonderful Life*, after its copyright expired. Anyone familiar with the commercial jingle for Quaker Puffed Wheat (and Rice) that begins "This is the cereal that's shot from guns" who has tried to listen to Tchaikovsky's *1812 Overture* has suffered the experience of having an exciting work permanently ruined. Similarly, Strauss's "Blue Danube Waltz" loses its magic to those familiar with the advertisement of the Rival Dog Food Company ("Give me Rival Dog Food, arf arf arf arf").

25. See, for example, the California Art Preservation Act, Cal. Civ. Code secs. 987–989 (Deering Supp. 1986); Mass. Gen. Laws Ann. chap. 231, sec. 85S (West Supp. 1985); N.Y. Arts and Cultural Affairs Law, secs. 11.01, 14.03 (McKinney Supp. 1986).

26. For a comprehensive discussion of state laws that promote nonpecuniary objectives, see Strauss (1960). In some circumstances, federal trademark law may also further some of the objectives underlying these rationales (15 U.S.C. sec. 1125[a]), which declares it a civil wrong to describe or represent falsely goods or services in interstate commerce.

27. 376 U.S. 225, 231–232 (1964). See also Compco Corp. v. Day-Brite Lighting, Inc., 376 U.S. 234 (1964), and Lear, Inc. v. Adkins, 395 U.S. 653 (1969).

28. See, for example, Aronson v. Quick Point Pencil Co., 440 U.S. 257 (1979) (requiring Quick Point to honor a contract requiring it to pay Aronson for use of an unpatentable design for a key ring). See also Goldstein v. California, 412 U.S. 546 (1973) (upholding a California statute protecting phonograph records despite the failure of copyright law to cover this subject matter).

29. 416 U.S. 470 (1974).

30. Harper & Row Publishers, Inc. v. Nation Enterprises, 471 U.S. 539, 555 (1985).

31. However, as other rationales become better incorporated into federal law, state protections premised on similar considerations may become more vulnerable to preemption claims.

32. 803 F.2d 1253 (2d Cir. 1986), *cert. denied,* 107 S.Ct. 2201 (1987).

33. See also Rosemont v. Random House, Inc., 366 F.2d 303 (2d Cir. 1966), *cert. denied,* 385 U.S. 1009 (1967); and Time, Inc. v. Bernard Geis Assoc., 293 F. Supp. 130 (S.D. N.Y. 1968).

34. 723 F.2d 195, 204 (2d Cir. 1983) (holding Ford's memoirs uncopyrightable), *rev'd,* 471 U.S. 539, 545–546 (1985) (upholding the copyrightability of the memoirs).

35. See Financial Information, Inc. v. Moody's Investors Services, Inc., 808 F.2d 204 (2d Cir. 1986) (daily bond cards describing daily trading information on municipal bonds not copyrightable); Eckes v. Card Prices Update, 736 F.2d 859 (2d Cir. 1984) (book listing all baseball cards manufactured 1909–1979 and their probable market prices held copyrightable); and Hoeling v. Universal City Studios, Inc., 618 F.2d 972 (2d Cir.), *cert. denied,* 449 U.S. 841 (1980) (research on burning of *Hindenberg,* including original theories on why incident occurred, not protectable).

36. Metropolitan Opera Ass'n., Inc. v. Wagner-Nichols Recorder Corp., 199 Misc. 786, 101 N.Y. 2d 483, *aff'd,* 279 App. Div. 632, 107 N.Y. 2d 795 (1950). See generally Gorman (1963).

37. Novel copyright theories have also, on occasion, proved successful, but these, too, turn on fact patterns that may not be shared across the spectrum of cases requiring protection. West Publishing co. v. Mead Data Central, Inc., 799 F.2d 1219 (8th Cir. 1986), *cert. denied,* 107 S.Ct. 962 (1987), is an example of a case in which a novel copyright theory worked. West, which publishes the decisions of most United States courts in hard copy and maintains a computerized data base of these same decisions, sued a firm running a rival computer data base. Although the cases themselves were in the public domain and were, therefore, not copyrightable, the Eighth Circuit strained to find a way to protect West, the original compiler. It therefore allowed Mead Data free use of the decisions, but enjoined it from noting the page numbers on which the cases appeared in West's publications. But since courts and legal publishers require citation to West page numbers (see, for example, *A Uniform System of Citation* [14th ed., 1986]), the decision protected West's market for hard copies of the decisions.

38. Cf. Diamond v. Diehr, 450 U.S. 175 (1981) (upholding patent on an industrial process that incorporated a computer program) with Parker v. Flook, 437 U.S. 584 (1978) (invalidating a patent on a computer program that was part of an industrial process).

39. Tie-outs restrict the licensee's right to sell products that compete with the patented product.

40. Tie-ins require licensees to buy from the patentee products used with or in the patented invention.

41. In addition to commenting on the No-Nos, Andewelt, who was speaking in his capacity as Antitrust Division deputy director of operations, expressed doubts over the department's longstanding opposition to patent protection for computer programs and its traditional hostility to trade secrecy laws.

Of course, the patentee and his licensee may together conspire to reduce the public welfare, but that problem can be handled with more finely tuned regulations than the No-Nos. See generally Baxter (1966).

42. 17 U.S.C. secs. 407, 408, 411.

43. 35 U.S.C. sec. 112. In cases in which the invention cannot be adequately described, deposit can also be required for patents (35 U.S.C. sec. 114).

44. 17 U.S.C. secs. 901–914 (Supp. II 1984).

45. Under 37 C.F.R. sec. 211.5, the applicant is permitted to preserve his trade secrets by withholding two to five layers of his chip design and blocking out sensitive portions of the ones that are revealed, so long as a major portion of the chip is revealed in the deposit.

46. See 37 C.F.R. sec. 202.20(c)(vii) (permitting deposit of only "identifying portions of the program," generally the first and last twenty-five pages).

47. See Berkey Photo v. Eastman Kodak Co., 603 F.2d 263, 281 (2d Cir. 1979) (refusing to require Kodak to disclose details of its 110 camera to film manufacturers), *cert. denied*, 444 U.S. 1093 (1980); see also ILC Peripherals Leasing Corp. v. IBM Corp., 458 F. Supp. 423, 436–437 (N.D. Cal. 1978) (holding that IBM is under no duty to disclose interface information to producers of peripherals), *aff'd sub nom.* Memorex Corp. v. IBM Corp., 636 F.2d 1188 (9th Cir. 1980).

48. See 29 C.F.R. sec. 1910.1200.

49. N.J. Stat. Ann. sec. 34: 5A-1–42.

50. But see Dow Chem. Co. v. United States, 476 U.S. 227 (1986) (sustaining, as against a trade secrecy challenge, the Environmental Protection Agency's right to overfly chemical plants to ascertain compliance with environmental regulations); Ruckelshaus v. Monsanto Co., 467 U.S. 986 (1984) (upholding data-sharing provisions of the Federal Insecticide, Fungicide and Rodenticide Act against trade secrecy challenge); and New Jersey Chamber of Commerce v. Hughey, 774 F.2d 587 (3d Cir. 1985) (sustaining New Jersey's right-to-know law).

51. In some circumstances, courts can also be called upon to reinterpret a statute to meet newly perceived problems. Other difficulties may be ameliorated through the actions of the government agency charged with administering the law. For example, the Copyright Office has recently promulgated a rule describing the circumstances in which it believes the copyright in a colorized motion picture should be recognized; see 52 Fed. Reg. 23691 (1987) (to be codified at 37 C.F.R. sec. 202.20(c)(2)(ii)).

52. See Public Law No. 98-417, codified at 35 U.S.C. sec. 156.

53. See H.R. 5536, 99th Cong., 2d sess., 1986, and H.R. 2482, 99th Cong., 2d sess., 1986.

54. See H.R. 1623 100th Cong., 1st sess., 1987.

55. See H.R. 4585 and H.R. 4808, 99th Cong., 2d sess., 1986.

56. See, for example, Medvedev (1969) (describing how Soviet horticultural research was affected by genetic theories that were more congenial to communism than was neo-Mendelism).

57. To some extent, this is a problem under existing intellectual property laws as well. If the invention is so advanced that it cannot be appreciated during the term of protection, the inventor will not be able to use the grant of exclusivity to capture the benefit that his invention will eventually generate. In addition, mavericks sometimes have difficulty convincing the Patent Office that their inventions work and are therefore eligible for patent protection; see Smith (1984) (describing an inventor's efforts to patent a generator that seemingly violates the second law of thermodynamics).

58. 17 U.S.C. sec. 906 permits another to reproduce a protected work without permission if the purpose is "teaching, analyzing, or evaluating the concepts or techniques" embodied in the work.

59. Indeed, a bill protecting the design of useful objects has already been offered in Congress; see H.R. 379, 100th Cong., 1st sess., 1987.

60. For an example of a business practice directly affected by the state of intellectual property law, see Davidson (this volume) who describes the effect of the decision in Apple v. Franklin on IBM.

61. See, for example, the International Convention for the Protection of Industrial Property (the Paris Convention), 20 March, 1883, 25 Stat. 1372, T.S. No. 37; as revised at Brussels, 14 December, 1900; at Washington, 2 June, 1911, 38 Stat. 1645, T.S. No. 579; at the Hague, 6 November, 1925, 47 Stat. 1978, T.S. No. 834, 74 L.N.T.S. 289; at London, 2 June 1934, 53 Stat. 1748, T.S. No. 941, 192 L.N.T.S. 17; at Lisbon, 31 October, 1958, [1962] 1 U.S.T. 1, T.I.A.S. No. 4931; and at Stockholm, 14 July, 1967, [1970] 2 U.S.T. 1583, T.I.A.S. No. 6923; the Universal Copyright Convention Rev., 24 July, 1971, 25 U.S.T. 1341, T.I.A.S. No. 6839.

62. See S. Rept. 100-66, 100th Cong., 1st sess., 1987; see also, Assistant Secretary and Commissioner of Patents and Trademarks, *Report on the Operation of the Int'l. Transitional Provisions of the Semiconductor Chip Protection Act of 1984* (7 November, 1986).

63. One could, of course, imagine other forms of research in this area. For example, the vitality of research and development in countries with very different protection schemes could be compared with each other; cf. Fox (1986) (comparing antitrust policy in the United States and Europe). I do not address the larger issue of what kinds of research should or could be undertaken. See also Griliches (1987).

64. Of course, some studies attempt to correct for this problem as well; see, for example, Winter (this volume).

References

American Law Institute. 1976. *Restatement (Second) of Torts.* St. Paul, Minn.: American Law Institute, sec. 652A.

Andewelt, R. 1985. "Antitrust Perspective on Intellectual Property Protection." *Patent, Trademark, and Copyright Journal* 30 (25 July): 319.

Assistant Secretary and Commissioner of Patents and Trademarks. 1986. *Report on the*

Operation of the Int'l. Transitional Provisions of the Semiconductor Chip Protection Act of 1984 (7 November).

Baxter, W. 1966. "Legal Restrictions on Exploitation of the Patent Monopoly: An Economic Analysis." *Yale Law Journal* 76: 267.

Bennetts, L. 1986. "'Colorizing' Film Classics: A Boom or a Bane?" *New York Times*, 5 August 1986, p. A-1, col. 3.

Dreyfuss, R. 1986. "Dethroning *Lear*: Licensee Estoppel and the Incentive to Innovate." *Virginia Law Review* 72: 677, 726-729.

Fox, E. 1986. "Monopolization and Dominance in the United States and the European Community: Efficiency, Opportunity, and Fairness." *Notre Dame Law Review* 61: 981.

Gordon, W. 1982. "Fair Use as Market Failure: A Structural and Economic Analysis of the *Betamax* Case and Its Predecessors." *Columbia Law Review* 82: 1600.

Gorman, R. 1963. "Copyright Protection for the Collection and Representation of Facts." *Harvard Law Review* 76: 1569.

Griliches, Z. 1987. "R&D and Productivity: Measurement Issues and Econometric Results." *Science* 237 (3 July): 31.

Hanley, R. 1987. "Fight Erupts on Baby M Book and Film Rights." *New York Times*, 26 March 1987, p. A-1, col. 2

Jewkes, J., D. Sawers, and R. Stillerman. 1969. *The Sources of Invention*. 2d ed. New York: W. W. Norton, pp. 189-192.

Kitch, E. W. 1977. "The Nature and Function of the Patent System." *Journal of Law and Economics* 20: 265, 276-279.

Kwall, R. 1983. "Is Independence Day Dawning for the Right of Publicity?" *U.C. Davis Law Review* 17: 191.

Lewin, T. 1982. "Whose Life Is It Anyway? Legally, It's Hard to Tell." *New York Times*, 21 November 1982, sec. 2, p. 1, col. 6.

Lipsky, A. B. 1981. "Current Antitrust Division Views on Patent Licensing Practices." *Antitrust Law Journal* 50: 515.

Machlup, F. 1958. *An Economic Review of the Patent System: Study No. 15 of the Subcommittee on Patents, Trademarks, and Copyrights, Senate Committee on the Judiciary*. 85th Cong., 2d sess., 22-25.

Medvedev, Z. 1969. *The Rise and Fall of T. D. Lysenko*. New York: Columbia University Press.

National Academy of Sciences. 1962. *The Role of Patents in Research*. National Academy of Sciences-National Research Council, pp. 22-23, 44, 48-49, 57.

Noyce, R. 1977. "Microelectronics." *Scientific American* 237: 63.

Prosser, W. 1971. *Handbook of the Law of Torts*. 4th ed. St. Paul, Minn.: West Publishing, sec. 117, p. 807.

Rule, C. 1986. "The Antitrust Implications of International Licensing: Analyzing Patent and Know-How Licenses." *Trade Regulation Reports* (Commerce Clearing House) 5 (21 October), par. 50, p. 482.

Scherer, F. 1980. *Industrial Market Structure and Economic Performance*. 2d ed. Boston: Houghton Mifflin, pp. 457-458.

Schneider, K. 1987. "Science Debates Using Tools to Redesign Life." *New York Times*, 8 June 1987, p. A-1, col. 2.

Smith, R. 1984. "An Endless Siege of Implausible Inventions." *Science* 226 (16 November): 817.

Strauss, W. 1960. *The Moral Right of the Author: Study No. 4 of the Subcommittee on Patents, Trademarks, and Copyrights, Senate Committee on the Judiciary*. 86th Cong., 1st sess., p. 141.

Wolfe, T. 1983. "The Tinkerings of Robert Noyce." *Esquire* 99 (December): 346.

2. Patents in Complex Contexts: Incentives and Effectiveness

Sidney G. Winter

Most of the discussion in this volume proceeds on the assumption that better protection of intellectual property is probably a good thing from an economic point of view. Questions have, of course, been raised about possible conflicts between economic incentives and other value systems that affect the advance of knowledge. In this chapter, I point out that the common assumption regarding the economic consequences of strengthened protection for intellectual property is also open to question. There is no theoretical presumption that these consequences are favorable. Generalization is genuinely hazardous, and an assessment of likely costs and benefits in a particular case requires a confrontation with complex empirical issues. As will become evident, I make no claim of originality in setting forth these propositions.

Following the review of these theoretical issues, I will present empirical evidence on the functioning of one major component of society's system for intellectual property protection: the patent system. This evidence is drawn from the Yale survey of R and D executives, a study conducted by Richard Levin in collaboration with Alvin Klevorick, Richard Nelson, and me. Two major results of the Yale study are highlighted here. The first is that, in general, the patent system does not play anything like a dominant role among the various mechanisms by which returns from innovation are captured. The second is that the patent system functions very differently in different manufacturing industries. A conceptual scheme that provides some insight into likely sources of these differences is also set forth.

Patents and Incentives: Theoretical Issues

Turning first to the theoretical issues, it should be acknowledged that at one time economists generally agreed on the desirability of strengthening intellectual property rights to promote innovation. In his landmark article of 1962, Kenneth Arrow provided a systematic account of the obstacles that impede the functioning of markets for information. He observed, among other things, that

> no amount of legal protection can make a thoroughly appropriable commodity of something so intangible as information. The very use of the information in any productive way is bound to reveal it, at least in part.
>
> The demand for information also has uncomfortable properties. . . . There is a fundamental paradox in the determination of demand for

> information; its value for the purchaser is not known until he has the information, but then he has in effect acquired it without cost. (1962, p. 615)

An alternative way of stating the last point is that it is difficult for the seller to certify, without some disclosure, his possession of information of a particular value to the potential purchaser.

Arrow's article also contained a simple (but seminal) formal model of process innovation. The message of this model was unambiguously that even an ideal system of patents (of infinite duration and costlessly enforceable) might well provide an "inadequate" incentive to invent. An inventor should realistically anticipate receiving something less than the gross social return from his or her inventive efforts. Arrow argued that an inventor would, therefore, abstain from making such efforts in cases where the expected costs of invention were large enough, compared to the social return, to make this shortfall of private returns critical. To the extent that any policy implication is inferable from this very simple economic model of invention, it is clearly in the direction of stronger protection for intellectual property rights, and perhaps other measures to reduce transaction costs in markets for such rights.

The general objection to this analysis is that the social return to an invention does not, in general, constitute an upper bound to the private rewards potentially capturable by the inventor—even when the economy's patent system is imperfect or nonexistent. Rather, the private return may consist substantially of transfers of returns ("rents" or "surpluses") from other actors in the system. Since this is the case, incentives for invention can be excessive in the sense that the resources absorbed in the inventive process may exceed the gross social return. The surplus from the inventor's private viewpoint (if any) would be made up by transfers from others. In such a case, strengthening of patent protection could exacerbate a preexisting tendency for excessive allocation of resources to certain types of inventive effort.

This possibility is obscured in the Arrow model in two different ways. First, the model presents a partial equilibrium analysis; price and quantity variations in a single market are explored on the assumption that prices in other markets are not significantly affected by the innovation. Here, as elsewhere in economic theory, partial equilibrium analysis yields only an approximation to general equilibrium results, and the approximation can be a very poor one. Hirshleifer (1971) observed that an inventor could, before disclosing the existence of his invention, invest in complementary assets whose prices would tend to increase as the invention was disclosed and implemented. The inventor's gain from the reevaluation of resource endowments induced by the improvement in society's technology depends on the endowment he or she holds, and bears no necessary relation to the social gains from the improvement. If the appropriate financial markets exist, the inventor might choose to go short in assets whose prices would be adversely affected by the invention, such as the stock of a corporation producing an existing alternative means to the same end.

It should be noted that in a general equilibrium context there may be no unidimensional magnitude representing social gains. One can only seek to identify who gains and who loses. It is theoretically possible that even in a competitive economy with *no* intellectual property protection, an improvement in technology could make everybody but the inventor worse off. (The inventor might gain from the reevaluation of his or her resource endowment.) If afforded additional institutional protection, the inventor may of course do even better at the expense of everyone else.

David Teece (1987) explores a closely related theme from a transaction cost perspective, relating appropriability to the transactional characteristics of the markets for complementary (or "co-specialized") assets. Teece argues that ease of access to such assets is a key consideration, perhaps *the* key consideration affecting appropriability among corporate inventor-innovators in the contemporary economy.[1] Corporate owners of inventions are, of course, likely to be much better positioned than individual inventors in terms of their access to assets complementary to their inventions. Even if they do not have market power in the markets for such assets, they stand to benefit from reevaluation effects.

The second way that the Arrow model obscures the possibility of excessive incentives has received a great deal more theoretical attention. The model does not allow for the fact that society might have more than one potential inventor for any particular invention. This generic issue was first explored in the literature of economic theory by Barzel (1968), who examined the timing of innovation on the assumption that the cost of making a given innovation is lower the later the innovation is made. He observed that under competitive "free access" conditions—in which there are no property rights in opportunities to make innovations—the tendency is for the innovation to be made at a date that is too early. The net social benefit would be larger if greater advantage were taken of the decline of costs over time. A large theoretical literature on "patent races" has explored variations of this theme under various assumptions about the structure of the market and of the innovative game.[2] In our book (1982), Nelson and I present a simulation study of innovation in an industry model; the results suggest rather strongly that unimpeded imitation need not yield inferior results from a social point of view.[3] Subsequent simulation work of my own, as yet unpublished, provides strong support for this suggestion.

These theoretical and simulation results all rest on some version of a fundamental assumption that society does not depend on the services of a unique individual for making a particular addition to society's store of knowledge. Instead, the superior creativity or diligence of an individual may be responsible for advancing the date at which the addition occurs. This issue is of sufficient importance to warrant a brief digression. It has been explored by historians and sociologists not merely with respect to inventions of direct economic utility (where the frequency of patent disputes involving priority questions might be considered adequate evidence by itself), but also with respect to intellectual achievements ranging through science, mathematics, philosophy, and beyond.

In a classic essay on the subject, sociologist Robert Merton observes that the phenomenon of multiple independent discoveries has *itself* been discovered and rediscovered numerous times, and has frequently been put forward as a notable and peculiar feature of the history of a unique subject matter: "This is an hypothesis confirmed by its own history" (Merton 1961, p. 352). Among the striking examples adduced in Merton's account is a *Times* of London story from 1865 observing "that the progress of mechanical discovery is constantly marked by the simultaneous revelation to many minds of the same method of overcoming some practical difficulty" (Merton 1957). An article entitled "Are Inventions Inevitable?" by W. F. Ogburn and D. S. Thomas appeared in 1922 and is commonly regarded by American sociologists as the locus classicus for a generally affirmative view on this question. Merton notes that the "fact and associated hypothesis of multiples" was also stated in that same year by authors in England, France, and Russia. Merton's own empirical investigation of the "fact of multiples" provides striking evidence in support of the hypothesis.[4]

Underlying the phenomenon of multiple independent discoveries is the cumulative social process of the advance of knowledge, a process that provides an expanding base—which might be termed the *common of the known*—from which individual investigators make their forays into the unknown. When the investigators are in close touch with one another and freely share their discoveries as they are made, the discoveries are no longer independent but are the product of collective effort. In pure science, the free sharing of emergent results is impeded by concern for recognition of priority, but the same concern spurs prompt publication of results that are in hand.[5] When advances have economic utility, it is ordinarily supposed that concerns with appropriating pecuniary rewards provide a strong inhibition to such sharing. There are, however, counterexamples to this supposition. Allen (1983) provides an account of the operation of the process of "collective invention" in the development of blast furnace technology in England's Cleveland district during the period from 1850 to 1875. He discusses the conditions under which producers have an incentive to engage in such a process, and essentially argues that reciprocity norms and prestige concerns akin to those leading to prompt publication in science provide an informal structure within which such gains can be realized. A broadly similar pattern, with formal cross-licensing arrangements supplementing the informal reciprocity norms and technical rivalry, is characteristic of the U.S. semiconductor industry today (Levin 1982; Rogers and Larsen 1984).

The phenomenon of collective invention, like that of multiple independent discoveries, is indicative of the important role that the common of public knowledge plays in the advance of knowledge. The patent-race models, showing how excessive incentives to invent can result because patenting enables private capture of part of the public domain, are addressing a very real issue. Of course, if something is in the public domain at the time of application, the patent will be denied for want of novelty in the invention. The more significant question is whether the invention would, absent the patent system, appear in the public domain within the useful life of the patent. The Patent and Trademark Office does not, and could not, screen applications on this basis.

In a provocative article, Edmund Kitch (1977) took Barzel's argument about the possibility of multiple inventors and essentially turned it on its head. Kitch's "prospect" interpretation of the patent system emphasizes that the system should be evaluated with respect to the effect of a patent on postaward resource allocation as well as on the preaward incentives to invent. To the extent that patents constitute an *early* delimitation of rights in opportunities for technological development—early in the stream of resource commitments required for commercially successful innovation—patents may provide the legal basis for more efficient organization of the development process. This argument certainly provides a rationale for stronger patent protection in some cases that might be misjudged from a narrow "incentive to invent" perspective. It is especially relevant to contemporary policy disputes about "orphan drugs" and about the licensing of patents for inventions resulting from government-funded R and D. As Kitch noted, however, the prospect approach does not entail the desirability of an indiscriminate "strengthening" of intellectual property rights protection. Rather, it adds another important dimension in which different situations can be usefully distinguished (Kitch 1977, esp. pp. 283–287).

The conclusion from all this work is a resounding, "It all depends." In particular, the desirability of strengthening or weakening invention incentives in a particular context depends on the existing balance (in that context) between (1) the joint effectiveness of a variety of means of appropriating returns, and (2) the extent to which the advances in question are actually a net contribution to society rather than a capture of wealth from the public domain.

The Effectiveness of Patents: Survey Evidence

As noted in the introduction, the empirical segment of this discussion draws upon data gathered in the Yale survey of R and D executives (Levin et al. 1987).

This survey obtained answers from R and D executives on a wide range of questions relating to the appropriability of gains from innovation and the sources of technological opportunity in the industries with which they are familiar. The particular data presented here represent a small segment of this large and complex data set. No attempt will be made here at formal statistical inference. It will be argued, however, that some tentative conclusions of clear relevance to the questions addressed in this volume leap to the eye, even from these limited descriptive data.

Respondents to the survey gave us the benefit of their judgments as experts on the situation prevailing in particular lines of business, rather than as sources of information on the activities of their particular companies. By addressing the respondents in this way we sought to circumvent the problem of company confidentiality, which is obviously a major roadblock to the development of microdata on the subject of intellectual property.[6] A far more difficult problem, which we did not circumvent, is that of developing a survey panel that reflects the existence of the large number of small entrepreneurial firms involved in innovation. The survey's coverage of the universe of R and D managers is quite good if the measure of coverage is the percentage of R and D spending whose management is included. It is much less adequate if measured by the number of firms or individuals involved.

We obtained a total of 650 responses representing 130 different lines of business. There were 18 lines of business for which we obtained 10 or more responses. This group includes most of the major R-and-D-performing industries; naturally, we tended to hear from a lot of R and D managers in areas where there were many to be heard from, and much R and D going on.[7]

Table 2.1 displays, for the eighteen-industry group, industry mean scores drawn from the first question in the survey. Like the majority of the survey questions, this question called for responses in the form of ratings on a seven-point Likert scale. As shown in Table 2.1, the question calls for effectiveness ratings for six alternative ways of protecting the gains from both process and product innovations. The mean ratings shown relate to patents to prevent duplication.

It is worth noting that the alternative—patents to secure royalty income—usually scores below patents to prevent duplication as a means of protecting innovative gains. In the group of industries shown in Table 2.1, royalties rank higher than preventing duplication in four of the eighteen industries in the case of process innovation, and in only two for product innovation. Furthermore, in samplewide averages, this means of protecting gains ranks last for new processes and next to last for new products (secrecy being lower). These results support the view that grave transactional difficulties limit the functioning of markets for information (Arrow 1962) and suggests that the patent system as an

Table 2.1. Effectiveness Ratings of Patents to Prevent Duplication

Line of business	New processes	New products
Pulp, paper, and paperboard (SIC 261, 262, 263)	2.58	3.33
Inorganic chemicals (SIC 2812, 2819)	4.59	5.24
Organic chemicals (SIC 286)	4.05	6.05
Pharmaceuticals (SIC 283)	4.88	6.53
Cosmetics (SIC 2844)	2.94	4.06
Plastic materials (SIC 2821)	4.58	5.42
Plastic products (SIC 307)	3.24	4.93
Petroleum refining (SIC 291)	4.90	4.33
Steel mill products (SIC 331)	3.50	5.10
Pumps and pumping equipment (SIC 3561)	3.18	4.36
Motors, generators, and controls (SIC 3621, 3622)	2.70	3.55
Computers (SIC 3573)	3.33	3.43
Communications equipment (SIC 3661, 3622)	3.13	3.65
Semiconductors (SIC 3674)	3.20	4.50
Motor vehicle parts (SIC 3714)	3.65	4.50
Aircraft and parts (SIC 3721, 3728)	3.14	3.79
Measuring devices (SIC 382)	3.60	3.89
Medical instruments (SIC 3841, 3842)	3.20	4.73

[QUESTION: In this line of business, how effective is each of the following means of capturing and protecting the competitive advantages of new or improved processes/products? (1) Patents to prevent competitors from duplicating the process/product, (2) Patents to secure royalty income, (3) Secrecy, (4) Lead time (being first with a new process/product), (5) Moving quickly down the learning curve, (6) Superior sales or service efforts.] Scale: 1 = "not at all effective," 4 = "moderately effective," 7 = "very effective." Mean scores for eighteen industries with ten or more respondents.

NOTE: The material in this table appeared in slightly different form in Levin et al. (1987), tables 2 and 5.

institution generally falls short of overcoming these difficulties. This has direct bearing on Kitch's "prospect" interpretation of patents, since that interpretation is founded on the claim that the assignment of such rights greatly facilitates the subsequent functioning of markets in the information in question.

A number of features of the results in Table 2.1 deserve comment. I will touch upon them only briefly here, to set the stage for my subsequent discussion of a conceptual scheme that suggests some of the sources of these differences.[8] Note first that, with the single exception of petroleum refining, patents are rated more effective in preventing product duplication than process duplication. More specifically, it is interesting to note that two-thirds of the scores in the product column are above the midpoint level of 4 ("moderately effective"), while roughly two-thirds of those in the process column are below that level.

The second observation is that there is a great deal of interindustry variation. At the high end there is the pharmaceutical industry. The fact that patents are rated highly effective there comes as no surprise, especially in the case of product patents. The significance of patents in that competitive arena is attested to both by the prior scholarly literature (for example, Taylor and Silberston 1973) and, from time to time, by the business section of the newspaper. Other chemical industries and petroleum refining also score high. What seems particularly noteworthy about the group of industries showing scores below 4 in one or both columns is the presence of some highly progressive, as well as R-and-D-intensive industries (for example, computers, semiconductors, and aircraft and parts). Of course, since there are alternative means of protecting gains from innovation that are regarded as more effective than patents, there is no paradox in the fact that patents are rated as relatively ineffective in some progressive, R-and-D-intensive industries.

What are the considerations that limit the effectiveness of patents? Table 2.2 provides a general view, based on the full sample, of the information the survey produced on this question. For both products and processes, it is clear that the fact that firms can invent around patents is the most important consideration by a substantial margin. The term *inventing around* covers a multitude of activities that may or may not be sins. At the benign extreme, this term may simply mean providing through independent inventive activity an effective functional substitute for the patented process or product, so that the patent does not block the achievement of some larger innovative goal. The knowledge borrowed from the prior inventor's contribution could be limited to, at most, the insight that the function in question is a useful one to perform. Nothing remotely approaching patent infringement may be involved. At the other extreme, however, the new solution skirts the edges of the existing patent's scope with just enough room to spare to make a successful infringement action unlikely—and the judgment about what is enough room may involve assessment not only of the legal scope of the existing patent, but also of the strategic stakes, resources, alertness, and litigation-proneness of the patent holder. In between these extremes there lies a broad interval where the borrowing from the prior inventor is very real but probably noninfringing under prevailing patent law.

Another significant feature of Table 2.2 is the large difference between the product and process columns for the two responses "Not readily patentable" and "Patents disclose too much . . . information." Given that patents on processes score lower in effectiveness than those on products, it is not surprising (but reassuring in terms of the coherence of the survey results) that all of the limits on patent effectiveness are seen as more significant for processes than for products. The difference is the largest for the two responses mentioned.

Figure 2.1 presents an attempt to explain some of our data and other observed features of the patent system by establishing taxonomic criteria that distinguish the sorts of knowledge which are good candidates for effective patent protection from those which are not.[9] More specifically, Figure 2.1 is intended to identify the sorts of advances in knowledge that can be protected by

Table 2.2. Limits on the Effectiveness of Patents

	New processes	New products
Not readily patentable	4.32	3.75
Patents are unlikely to be held valid if challenged	4.18	3.92
Firms do not attempt to enforce patents	4.29	3.84
Competitors can legally "invent around" patents	5.49	5.09
Technology is moving so fast that patents are irrelevant	3.40	3.34
Patents disclose too much proprietary information	4.19	3.65
Licensing is required by court decisions or decrees	2.96	2.79
Firms participate in cross-licensing agreements with competitors	3.08	2.93

[QUESTION: To what extent does each of the following considerations limit the effectiveness of patents as a means of protecting the competitive advantages of new or improved production processes/products in this line of business?] Scale: 1 = "does not limit effectiveness of patents," 4 = "moderately important limit on effectiveness of patents," 7 = "very important limit on effectiveness of patents." Mean scores, full sample.

NOTE: The material in this table appeared in slightly different form in Levin et al. (1987), tables 2 and 5.

Figure 2.1. Conditions for Strong Individual Patents

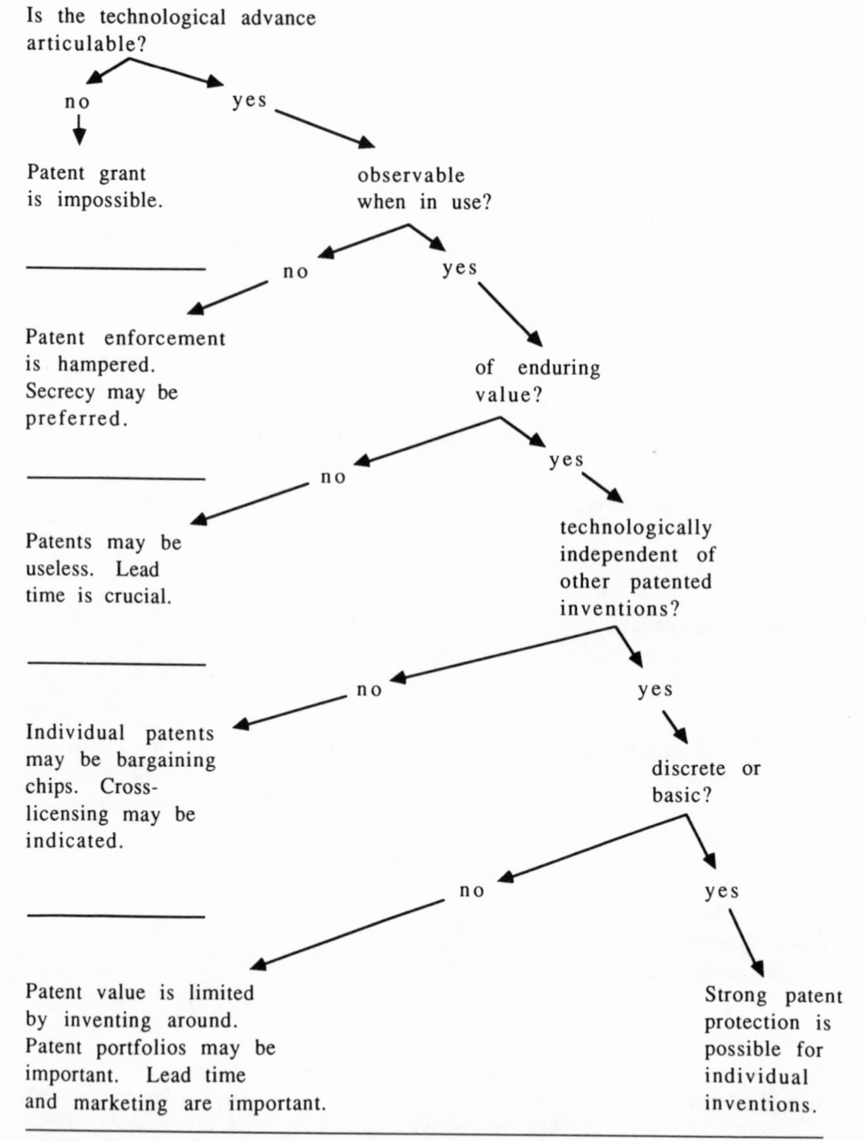

NOTE: The material in this figure appeared in slightly different form in Levin et al. (1987), tables 2 and 5.

strong individual patents. This is conceptually very close to the question on patent effectiveness posed in our survey, although we did not ask respondents to attempt the difficult task of assessing the typical effectiveness of individual patents in isolation. It is not the same issue as the *importance* of patents, since patent portfolios can be important even where individual patents are weak. Furthermore, the importance of patents is presumably reduced when alternative means of protecting the innovation are a close substitute for patent protection.

Most emphatically, the question of whether there are strong individual patents is not the same as the question of whether patents are numerous, in relation to sales or by some other measure, in a particular line of business. Consider an invention that meets standards of patentability as determined by the Patent and Trademark Office. The question of whether a patent is obtained presumably depends upon whether the benefits anticipated from having it surpass a threshold set by the costs of filing. On the other hand, whether a patent is "strong" is defined here as a matter of whether its validity would be upheld under determined attack in the courts, and whether the patent is also relatively invulnerable to noninfringing "inventing around." In short, a strong individual patent is one that functions like the patents in most economic models of patenting: it fully protects the economic returns on the knowledge to which it relates. It is well known that most patents are not strong.

In the tree structure of Figure 2.1, the vertical arrangement of the forks represents an attempt to reflect this important distinction between whether there are patents and whether there are strong individual patents. The "no" branches at the top of the page lead to outcomes involving no patents or few patents, while at the bottom of the page a "no" is compatible with a large volume of patenting, but with individual patents that are not strong.

At the uppermost fork, for example, the question is whether the knowledge advance in question is tacit or articulable. A great deal of economically significant process knowledge resides in individual skills and organizational routines that are either inherently tacit or are not articulable within the constraints of a patent application.[10] Improvements in this sort of knowledge can occur, at the individual level, without conscious awareness of how improvement has been achieved or perhaps even of the fact that improvement has occurred. Similarly, coordination improvements in organizations may occur for reasons that are imperfectly and diversely understood by the participants therein. Knowledge advances of these types fall outside the realm of statutory subject matter for patents. Furthermore, the requirement that a patent application contain "a written description of the invention . . . in such full, clear and concise terms as to enable any person skilled in the art to which it pertains . . . to make and use the same" (35 U.S.C. sec. 112) is an extremely demanding requirement when even moderately severe difficulties in articulation are present. In effect, it demands that the application be a complete instruction book for a "person skilled in the art." The fact that instruction books are not generally an efficient means of conveying improvements in skill is resoundingly attested by the fact that how-to books have not yet rendered firsthand instruction obsolete.[11]

Assuming that the advance is sufficiently articulable to be eligible for patenting, the next question is whether it is "observable when in use." The issue here is *reverse engineering*. In its narrow meaning, this term relates to situations in which the technological advance is embodied in a product that can be purchased, taken back to a laboratory, dismantled, and scrutinized in minute detail. Under these circumstances, it is a rare novelty whose secrets will not yield to determined attack. A patent application then discloses little that a competitor cannot discover anyway, although later and at greater expense. At the opposite extreme is a process improvement that is articulable (and hence potentially patentable), but that can be kept secret through ordinary steps, such as placing the process off limits to nonemployees and then limiting in various ways the security hazard inherent in the access provided to employees. This type of situation—illustrating the "no" branch at the second fork of Figure 2.1—involves a dual inhibition to patenting. The patent application discloses information that could be kept secret, and if a patent is obtained it will be difficult to enforce because infringement cannot be observed.

The contrast between product improvements that can be reverse engineered and process improvements that can be kept secret provides a plausible hypothesis to explain many of the differences between product and process responses to the appropriability questions in the Yale survey. This contrast is too sharp, however, to stand as an unqualified generalization about the distinction between product and process improvements. First, reverse engineering may be conceived more broadly than is suggested above. It can include all of the activities through which valuable inferences regarding the key elements of novelty in an invention can, by various sorts of observation, be derived. Second, the extreme case of easily achieved process security described in the contrast above is characteristic of few, if any, real situations.

Reverse engineering in the broad sense is often possible for processes—some helpful clues, at least, are obtainable by an external investigator with appropriate expertise. And on the other hand, product improvements are not universally vulnerable to reverse engineering to the narrower sense—although this approximation to reality is much closer than the one regarding processes.[12] Taken together, these qualifications indicate that "observable when in use" is more a matter of degree than a strict dichotomy (a point that applies to every fork of the tree). They also indicate that observability is not fully aligned with the product/process distinction. To take one example, the device of using a single queue when there are multiple servers was a process innovation of the 1970s in banking and other industries, although one that would obviously be impossible to keep secret.

At the next fork the question posed is whether the advance is of enduring value. The "no" branch corresponds, of course, to the fifth line of Table 2.2: "Technology is moving so fast that patents are irrelevant." In the most common and simplest interpretation of this situation, the problem is the delay between the time of application and the time the patent is issued. A recent story in the *Wall Street Journal* remarked that "nasty patent fights are rare in high technology, where the market can bypass an innovation by the time a patent is granted" (Kneale 1987). Taken literally, this statement suggests that inventions go unpatented because the patents would be literally valueless when issued. Because the invention is not legally protected until the patent is issued (and this may take years), it is certainly possible that this time lapse could render a patent valueless. It is also clear that the prospect of a patent becoming valueless (due to obsolescence) within a few years after issuance detracts from, even if it does not eliminate, its value at the time of issuance.

I would speculate, however, that the important question here is that of the patent's value in relation to the costs of applying for it—where those costs include the time and attention of technical personnel who could, instead, be producing more advances. More broadly, perhaps a corporate culture that is strongly oriented to legal protection of the gains from past achievements is poorly suited to high-tech environments. As Figure 2.1 indicates, what is crucial when technology is moving fast is lead time—the ability to embody the technology in well-designed products or service systems and produce and market them before rivals can. An intense and sustained organizational focus on this goal may involve a sacrifice of attention to the rearguard action of patent protection.

The survey results provide some statistical evidence in support of this suggestion. The simple cross-industry correlation between the two limits to patent effectiveness, "Firms do not . . . enforce patents" and "Technology is moving so fast that patents are irrelevant," is 0.18 for process innovations and 0.23 for product innovations (significant respectively at 5 percent and 1 percent levels, $N = 130$).

The next fork of the tree poses the question of whether the advance under consideration is independent of other patented inventions. The significance of this question derives from the following important point about the patent system:

> A patent does *not* give the inventor the positive right to make, use, or sell his own invention, but rather grants to the inventor the negative right *to exclude others* from making, using, or selling the invention.

> Whether an inventor may have received a patent on an invention has nothing to do with whether the inventor himself can make, use, or sell the invention. Indeed, in making, using, or selling his own invention, the inventor may find that he infringes the patent rights of others. (Burge 1984, p. 27, emphasis in original)

For the production of complex systems embodying numerous patented inventions held by different parties, there are three main institutional forms: (1) ad hoc licensing agreements entered into for the specific purpose of legally enabling the production of a particular system, (2) broad cross-licensing agreements that give a number of parties full access to a substantial technology pool, and (3) system development that simply proceeds within a particular firm that considers itself to be in a strong patent position and is prepared to deal with possible infringement suits as they arise. Only the first of these arrangements may give the individual patent "its due" in that a royalty rate specific to the patent may be determined as part of the overall agreement. In the other cases, the individual patent is simply one bargaining chip among many, serving to augment slightly the willingness of other firms to participate in cross-licensing, or to improve somewhat the prospects of its owner in settlements or final judgments in patent infringement suits.

On the basis of interviews with corporate patent attorneys in the semiconductor field, Eric von Hippel set forth the following incisive characterization of this sort of situation:

> Firm A's corporate patent department will wait to be notified by attorneys from firm B that it is suspected that A's activities are infringing B's patents. Since possibly germane patents and their associated claims are so numerous, it is in practice usually impossible for firm A—or firm B—to evaluate firm B's claims on their merits. Firm A therefore responds—and this is the defensive value of patents in the industry—by sending firm B "a pound or two" of its [possibly] germane patents, with the suggestion that, while it is quite sure it is not infringing B, its examination shows that B is in fact probably infringing A. The usual result is cross-licensing, with a modest fee possibly being paid by one side or the other. (Von Hippel 1982, p. 102)

Patents that are bartered by the pound clearly are not strong individual patents.

At the bottom fork in Figure 2.1, the issue is vulnerability to inventing around. This issue is complex, and the diversity of situations presented in the patent system is obviously enormous. A systematic approach to grading inventions on this dimension of vulnerability would be a useful thing to have, but at this point I can offer only a few preliminary observations on the problem. First, the fact that "ideas" are not patentable per se is key. It means that the cognitive processes of aspiring inventors can freely manipulate ideas regardless of origin—and, in particular, regardless of their being evoked by observation of patented inventions (or patent applications) in the same field in which the inventor is working. Patent law does not completely bar the door to this sort of borrowing, nor should it. One implication is that a degree of vulnerability to inventing around is inevitable.

Second, it seems clear that what are called basic patents are less vulnerable to being invented around than others. This may be in part a tautology; we ascribe basicness where we observe invulnerability. Typically, however, a basic patent is distinguished operationally by its chronological primacy in a stream of related inventions and by the absence of a closely related prior art, whereas applications for later patents in the stream cite the basic patent as prior art. It is worth noting that basic patents are not necessarily valuable just because they are strong. Historical examples illustrate a wide variety with respect to the chronologies of patenting and commercial success. Sometimes the basic patent has long since expired before significant commercial success is achieved; sometimes a viable company or a new industry is founded shortly after (or even before) the patent on the key invention has been issued (Kitch 1977, p. 272).

Finally, intrinsic features of different areas of technological art seem to facilitate or frustrate efforts to invent around a patent. In the case of mechanical inventions, an idea borrowed from an existing patented invention may be reembodied in an alternative device that involves a host of modifications of configuration and materials, and perhaps novel solutions to ancillary difficulties associated with the idea. Whether such a device actually infringes the original patent will certainly depend in part on the skill of the patent attorneys who drew up the original application. Equally significant is the fact that the outcome of an infringement suit will also depend upon the high-priced skills of attorneys. The question is not just one of the trial result, but of how much it would cost and who would pay the fees. As was noted above, success in inventing around a patent may be represented by an achievement that falls well short of being *plainly* noninfringing.

The pharmaceuticals industry provides a strikingly different illustration of how intrinsic features of the underlying art interact with the patent system and other institutional arrangements; in this case, the interaction is one that tends to produce highly effective patents. Fundamental to this result is the fact that a particular chemical molecule is an entity that is discrete and describable in precise terms. The options available in the case of mechanical inventions, where it is possible to borrow a basic idea and embody it in an invention with different materials or configuration, are largely absent from the chemicals sector. Ways of modifying the molecule that are reliably inconsequential may be available, but many of these can be anticipated and covered in the original patent. Inventing around a drug patent is generally difficult.

The regulatory process affects patent strength in a significant way. The FDA approval that is required before a drug can be marketed relates to a specific molecule; in fact it relates to one or more specific modes of therapeutic use of a specific molecule. Such approval does not extend to the product of an imitator who, unwilling to take a license under a patent still in force, seeks to invent around the patent. As Grabowski and Vernon concisely state:

> In contrast to imitative products involving already approved substances by the FDA (i.e., generic equivalents), chemically differentiated products must undergo full-scale reviews of safety and efficacy by the FDA. Hence, these drugs must be tested on the same scale as all previously approved products. Therefore, under current regulatory conditions, the imitating firm is faced with several million dollars in development costs and several years in lag time before chemically distinct follow-on drugs can be marketed as approved new drugs. (1982, pp. 294–295)

This regulatory system is a powerful complement to the patent system in that it provides specific protection for patent rights precisely where they tend to be weakest, in their vulnerability to inventing around.

There are further twists and turns to the story of pharmaceuticals patents. Part of the "discreteness" of a new drug derives from the weakness of the theory connecting chemical properties to therapeutic effect; this makes it hard to borrow the idea underlying the therapeutic effect and achieve it in an alternative way. As noted by Grabowski and Vernon, however, these theoretical connections are rapidly being strengthened, and the character of pharmaceutical R and D is changing accordingly (1982, esp. pp. 294, 303, 308–309).

Also, when a drug patent expires, the established brand name of the original producer may make it possible to sustain a price premium over the generic equivalents that come on to the market. (Since the producers of generic equivalents do not have to provide *therapeutic* equivalence to the FDA regulators, such a premium may in some cases be justified.) Oster (1982) discusses the ways in which different groups in the industry have interacted with, and in, the regulatory process on the question of generic substitution. Such problems as the length of the effective patent life and the status of orphan drugs are also important areas of policy discussion relating to intellectual property in the pharmaceuticals industry.[13]

Conclusion

A simple conclusion emerges from the discussion above: These are not simple matters. Theoretical problems involving the relationship of intellectual property rights to incentives for economic progress are complex and multifaceted, even at the high levels of abstraction at which they have been addressed. A great deal remains to be done in the analysis of theoretical models seeking to allow for (1) richer characterization of technological opportunity, (2) the important role played in appropriability regimes by considerations other than intellectual property rights, and (3) implications of the time sequencing of investment commitments in the linked processes of invention and innovation. On the empirical side, there is plenty of evidence for the general proposition that industrial contexts are extremely diverse in ways relevant to intellectual property policy, and for the specific proposition that patents do not always, or even generally, play a key role.

Regarding the policy issues, it is tempting to conclude that a discriminating approach is called for, and that such an approach needs to be grounded in thorough analysis of particular situations. An attempt in this direction is certainly likely to produce more sensible results than a rhetorical battle fought with broad generalizations and simplistic analyses. In addition, such an attempt would afford some protection against the dangerous error of presuming that proposed changes in general policy will be nonselective in their effect, merely on the grounds that the specific industries affected have not been named or identified (Levin 1985).

Even the call for a more discriminating approach cannot, however, be advanced without qualification. The institutions that are in place, such as the Patent and Trademark Office, have a difficult enough time implementing relatively blunt policy tools, and significant administrative costs are incurred in the process. To suggest that something more subtle should be attempted is to raise obviously serious questions of administrative feasibility and cost. An ideal policy would strike an appropriate balance between these considerations and the benefits of a discriminating approach to a complex environment. It is clear that we are a long way from knowing how to design an ideal policy.

At a practical level, I confess that any policy stance I might personally take would be heavily influenced by the axiom colloquially stated as "If it ain't broke, don't fix it." If something is (allegedly) "broke," first attention should go to who says so, how long it has been that way, and what caused the alleged breakage. I would be wary of proposals for radical institutional change promoted by special interests and general rhetoric.[14]

Notes

Acknowledgment: Financial support from the National Science Foundation, Division of Policy Research and Analysis, and from the Sloan Foundation is gratefully acknowledged.

1. Insofar as *innovation* (the economic introduction of a new process or product)

involves costly problem-solving activity additional to that involved in the underlying *invention*, the information-theoretic economics of invention and innovation are basically the same. For most purposes in the present discussion the two terms are interchangeable. Where this is not the case, the specific meaning should be clear from the context.

2. For a particularly informative example of this genre, see Gilbert and Newberry (1982).

3. See especially the results in the "science based, easy imitation" condition of the experiments reported in chap. 15.

4. See his discussion of an intensive analysis of 264 examples of multiple discovery (Merton 1961, pp. 364–365). See also S. C. Gilfillan's review of the history of inventions in shipbuilding, which leads him to the following observation: "We cannot find a single example to offer, but if there were some desired discovery so difficult, so recondite that only five men in the world had the intellect and the preoccupations which would lead them to it—not one of those geniuses would be necessary for the invention, since any of the remaining four would find it instead, if one were removed" (Gilfillan 1935, p. 74).

5. The conflict in the institution of science between the value of originality (hence priority) and other values such as "communism of intellectual property" and, especially, humility has been addressed by Merton (1957, 1963).

6. Even though the questions do not directly relate to individual companies, we have pledged confidentiality to our respondents and will not release the data in forms that would permit individual responses to be identified or approximated.

7. For details on the design of the total survey and an overview of the results relating to appropriability, see Levin et al. (1987).

8. See Winter (1987) for a more extensive commentary on Table 2.1.

9. I am indebted to Richard Levin for suggesting some improvements in Figure 2.1 relative to the version presented at the conference. He is absolved, however, from responsibility for the final result.

10. See Nelson and Winter (1982, chaps. 4 and 5) and Winter (1987) for discussion of the sources and consequences of tacit knowledge in individual and organizational performance.

11. The exclusion from protection of processes that require "a mental step to be performed" is another quite explicit bar to the protection of tacit knowledge.

12. For example, producers who sell highly customized products in narrow markets are multiply advantaged in protecting against reverse engineering. Not all the secrets relevant to the product class are at risk in every installation; their customers are known to them and can be contracted with on the matter; and access needed for servicing and repair may facilitate enforcement of such an agreement.

13. See Grabowski and Vernon (1982, esp. pp. 344–354) for a summary discussion of these issues and others.

14. Although cast in less graceful language, this conclusion closely follows that of Machlup (1958).

References

Allen, R. C. 1983. "Collective Invention." *Journal of Economic Behavior and Organization* 4: 1–24.

Arrow, K. J. 1962. "Economic Welfare and the Allocation of Resources for Invention." In *The Rate and Direction of Inventive Activity*, edited by R. R. Nelson, pp. 609–624. Princeton: Princeton University Press.

Barzel, Y. 1968. "Optimal Timing of Innovations." *Review of Economics and Statistics* 50: 348–355.

Burge, D. A. 1984. *Patent and Trademark Tactics and Practice*. 2d ed. New York: Wiley-Interscience.

Gilbert, R. J., and D.M.G. Newberry. 1982. "Preemptive Patenting and the Persistence of Monopoly." *American Economic Review* 72: 514–526.

Gilfillan, S. C. 1935. *The Sociology of Invention*. Cambridge: MIT Press.

Grabowski, H. G., and J. M. Vernon. 1982. "The Pharmaceutical Industry." In *Government and Technical Progress: A Cross-Industry Analysis*, edited by R. R. Nelson, pp. 283–360. New York: Pergamon.

Hirshleifer, J. 1971. "The Private and Social Value of Information and the Reward to Inventive Activity." *American Economic Review* 61: 561–574.

Kitch, E. W. 1977. "The Nature and Function of the Patent System." *Journal of Law and Economics* 20: 265–290.

Kneale, D. 1987. "Patent Disputes Could Affect Disk-Drive, Network Markets." *Wall Street Journal*, 13 February 1987, p. 27.

Levin, R. 1982. "The Semiconductor Industry." In *Government and Technical Progress: A Cross-Industry Analysis*, edited by R. R. Nelson, pp. 9–100. New York: Pergamon.

______. 1985. "Patents in Perspective." *Antitrust Law Journal* 53: 519–522.

Levin, R., A. K. Klevorick, R. R. Nelson, and S. G. Winter. 1987. "Survey Research on R&D Appropriability and Technological Opportunity." *Brookings Papers on Economic Activity* 1987: 783–820.

Machlup, F. 1958. *An Economic Review of the Patent System: Study No. 15 of the Subcommittee on Patents, Trademarks, and Copyrights, Senate Committee on the Judiciary*. 85th Cong., 2d sess.

Merton, R. K. 1957. "Priorities in Scientific Discovery." In *The Sociology of Science: Theoretical and Empirical Investigations*, pp. 286–324. Chicago: University of Chicago Press, 1973.

______. 1961. "Singletons and Multiples in Science." In *The Sociology of Science: Theoretical and Empirical Investigations*, pp. 343–370. Chicago: University of Chicago Press, 1973.

______. 1963. "The Ambivalence of Scientists." In *The Sociology of Science: Theoretical and Empirical Investigations*, pp. 383–412. Chicago: University of Chicago Press, 1973.

Nelson, R. R., and S. G. Winter. 1982. *An Evolutionary Theory of Economic Change*. Cambridge: Harvard University Press.

Ogburn, W. F., and D. S. Thomas. 1922. "Are Inventions Inevitable?" *Political Science Quarterly* 37: 83–98.

Oster, S. 1982. "The Strategic Use of Regulatory Investment by Industry Sub-groups." *Economic Inquiry* 20: 604–618.

Rogers, E. M., and J. K. Larsen. 1984. *Silicon Valley Fever: Growth of High-Technology Culture*. New York: Basic Books.

Taylor, C. T., and Z. A. Silberston. 1973. *The Economic Impact of the Patent System: The British Experience*. Cambridge: Cambridge University Press.

Teece, D. J. 1987. "Profiting from Technological Innovation: Implications for Integration, Collaboration, Licensing and Public Policy." In *The Competitive Challenge: Strategies for Industrial Innovation and Renewal*, edited by D. J. Teece. Cambridge, Mass.: Ballinger.

Von Hippel, E. 1982. "Appropriability of Innovation Benefit as a Predictor of the Source of Innovation." *Research Policy* 11: 95–115.

Winter, S. G. 1987. "Knowledge and Competence as Strategic Assets." In *The Competitive Challenge: Strategies for Industrial Innovation and Renewal*, edited by D. J. Teece. Cambridge, Mass.: Ballinger.

3. Recent Changes in the Patent Law That Affect Inventorship and Ownership

Patrick D. Kelly

Since 1984 the U.S. patent laws have gone through major changes that directly affect the laws of inventorship, rendering the patent system much more flexible and better suited to handle the complexities and the shades of gray that exist in the real world of innovation. Among other things, the new laws give researchers at all levels (including research managers and technicians) a better chance to obtain recognition as co-inventors of patents.

The Law of Inventorship

In order to be an inventor, someone must contribute to the "conception" of the invention a complete and functional idea of how the invention will work. One must do more than merely envision a final result, such as a "cure for cancer" or "a plasmid containing the gene for Cureall." The inventor must understand a way to achieve that result; he or she must be able to describe the invention in such a way that other skilled researchers can carry it out successfully without "undue experimentation," and every component or reagent necessary to make the invention must be available.

As an example, these requirements explain why the science fiction writer Arthur C. Clarke could not get a patent on his idea for geosynchronous satellites. He first published his idea in the 1940s—long before anyone had the ability to get rockets and satellites into space. Even though the theoretical principles of rocketry were known, a tremendous amount of work still had to be done before rockets worked, and even more remained before satellites could be positioned and oriented accurately. Clarke's proposal was a statement of a goal, not a conception of an invention.

In order for an invention to be complete, the inventor must reduce the invention to practice. Usually, this means actually creating a prototype of the invention. It might be a working model (in the case of electrical and mechanical inventions), a small quantity of a compound in the case of chemical inventions (with analytical data proving that the compound exists and with experimental data showing that it has a valid use), or a plasmid or culture of cells in a genetic-engineering invention.

In some cases (especially in the mechanical field, where components are predictable), an inventor can reduce an invention to practice without actually having to build or test it, if he or she can describe the invention in enough detail to teach "anyone with ordinary skill in the art" (35 U.S.C. sec. 112) how to make and use it. However, an inventor who follows this route assumes a risk. If, through oversight, an important element is left out of the description, the patent might not be issued or it might not be enforceable. Although prophetic patents (sometimes called paper patents) can be used defensively (that is, they can protect a company's right to operate in a certain area), it is usually better for the inventor actually to build and test an invention before a patent application is filed.

If a technician, working under the direction of a senior scientist, professor, or research manager, does the actual work of reducing an invention to practice, a question arises as to whether the technician is a co-inventor. The answer depends on whether the technician contributed to the conception of the invention. At one end of the spectrum, if the technician merely carried out clear and explicit instructions, he or she is not a co-inventor. At the other extreme, if the supervisor merely stated a goal and left it to the technician to figure out how to reach that goal, and if the technician had to develop a new technique that was not "obvious to those skilled in the art" in order to reach that goal, then the technician is a co-inventor.

Many, many cases of real-world inventorship fall somewhere between those two extremes. In particular, research managers and technicians fall into the gray area almost constantly. But the patent law does not recognize shades of gray; it recognizes only black and white. Someone either is or is not an inventor of a patent. There is no halfway category of recognition short of inventorship, so a patent attorney has no choice but to force the complexities of real life to fit into either of two simplistic categories.

That type of decision can lead to badly hurt feelings and bitter unhappiness. Being recognized as co-inventor of a patent carries credit and prestige, and researchers who are omitted may be intensely disappointed and unhappy (especially early in their careers, when they are struggling to build up a record of accomplishment). There is no way to avoid tough inventorship decisions, but those decisions require a sensitive awareness of their human costs, which can include hard feelings and jealousy, badly damaged enthusiasm, disruption of the working relations in a lab, and lower productivity not just from the offended employee, but from everyone who depends on his or her work. It is also worth recognizing that the problems generated by failure to receive fair credit for an important contribution often do not go away in a few weeks. They can often set the tone for an entire career.

The worst problems arose before the patent law was changed in 1984. As described below, the revised law of inventorship is far more flexible. It minimizes the unfairness that comes from forcing real-world facts into a simplistic black-or-white system.

Regrettably, patents do not even contain a list of acknowledgments, the way most scientific articles do. I myself have tried to include a list of acknowledgments in patents when particular credit was due, but each time other patent attorneys insisted that such a gesture would invite trouble. I yielded, but I still think it is unfortunate that patents do not recognize noninventors who contributed substantially to an invention.

The Law of Patent Ownership

If a patent is invented by more than one person, the ownership usually falls into one of two categories. If all of the inventors assign their rights to a company, a university, or some other entity (called the *assignee*), then the assignee owns the entire patent. That is the simple case.

If the co-inventors retain their rights, each co-inventor owns an "undivided interest." The interest is called undivided since each inventor has the right to license others to use the entire patent. This is comparable to two renters sharing an apartment; both of them are entitled to use the entire kitchen, the entire bathroom, the entire living room, and so on. If they want to split up the bedrooms and take one each, they can do so, but only if they both agree.

In the absence of a written contract that divides up the rights to an unassigned patent, each co-inventor is entitled to license any number of companies to use the patent and all of its claims. However, each inventor (and his or her potential licensees) realizes that every other inventor has that same right. Therefore, a license from one co-inventor acts like a nonexclusive license, which is usually not very valuable. That gives co-inventors a powerful incentive to reach an agreement among themselves if they want to commercialize the invention.

Because every nonassigning co-inventor can license anyone to use the entire patent, corporations are very reluctant to list anyone outside of the company as a co-inventor of a patent. If an outside consultant (such as a university professor) is involved in a company project that leads to an invention, the patent attorney will know whether the company has a right of assignment to inventions by that consultant involving that project. If so, co-inventorship is not a problem; the company will simply ask the consultant to assign his rights to that patent to the company, and the consultant must do so under the research contract.

But a problem can arise if the consultant is not required to assign to the company any invention that results from the project. In that case, logic suggests that the company's patent attorney will be under pressure to decide that the only inventors were company employees. If an outsider is listed as a co-inventor, the patent is in danger of becoming worthless to the company except for defensive purposes. This type of problem can usually be worked out through a contract where the consultant (or his university) grants an exclusive license to the company. The situation can become difficult, however, and companies do well to obtain a right of assignment before signing any consulting contract.

The situation is even more complex if the professor's consulting is in a research area that is also supported by federal funds. The law states that a nonprofit organization such as a university will own any patent on an invention that is either "conceived or first actually reduced to practice" in the course of work that receives federal funding (35 U.S.C. sec. 201[e]). Suppose a chemical company thinks of a medical use for a new drug created by one of its scientists. Because the company does not maintain clinical facilities, it asks a professor at a medical school to supervise some clinical tests on the new drug. If the clinical tests are commingled with other federally funded research at the medical school, and if the clinical tests are considered the first reduction to practice of the idea, then the law might assign the patent to the medical school. To avoid such problems, a research agreement can include a provision requiring company-sponsored research to be separated from any federally funded research. Otherwise, the medical school may get the patent (or at least part of it), and the company must request an exclusive license from the school.

It is possible to change the inventorship of a patent. If a consultant (or anyone else, such as a former graduate student who no longer works with a certain professor) learns that a patent application was filed or a patent was issued on an invention from which he or she was unfairly omitted, the consultant may contact the company. It may be that the omission was an honest error, because the company's patent attorney did not have time to check out every potential claim by every potential co-inventor. (There is tremendous time pressure to get applications filed as soon as possible.) If an agreement is not reached after an initial discussion, generally the best course is to have an independent patent attorney study the facts and decide who should be listed as an inventor.

The Law of Co-Inventorship before 1984

Under the law that existed before November 1984, every co-inventor listed on a patent application had to be a co-inventor of every claim. This followed from the notion that a patent application could claim only one invention. That notion is simplistic, and it is often manipulated or ignored to achieve practical results.

For example, if Professor Smith invents a new plasmid, her patent application will claim the plasmid and a culture of cells containing the plasmid, even though a plasmid and a culture of cells are clearly different things. Suppose that the invention is a nonpapilloma plasmid for transforming mammalian cells. To get that plasmid into mammalian cells, she may have used a technique developed by Mr. Jones, who works down the hall. The most important claims in the application will read something like

Claim 1: A plasmid with certain characteristics.

Claim 2: A mammalian cell transformed by the plasmid of Claim 1.

Obviously, Jones did not contribute to Claim 1. Under the old law (before 1984), he could not be a co-inventor of a patent application covering that plasmid. That puts Smith and her university in a dilemma: Should the university pay for two applications, one by Smith covering the plasmid, and a second application by Smith and Jones covering the cells? Since each application involves expensive filing and attorney fees, universities and small companies are under pressure to file only one if they can get away with it. Under that pressure, could the patent attorney make a good argument that Jones's contribution in creating the cells did not rise to the level of inventorship?

Decisions like that can be made without much difficulty in the comfort of a quiet room, but they can unravel with disastrous results if an opposing attorney discovers them later and uses them as a way to cause trouble. If Jones was disappointed at being left off the patent, the application could become a time bomb. The 1984 law changed all that.

The 1984 Changes

On 8 November, 1984, Public Law No. 98-622 took effect. It modified several sections of the patent statute, including 35 U.S.C. sec. 116, which controls inventorship. The revised section now reads, in part: "Inventors may apply for a patent jointly even though (1) they did not physically work together or at the same time, (2) each did not make the same type or amount of contribution, or (3) each did not make a contribution to the subject matter of every claim of the patent." Continuing with the Smith and Jones example above, a single application can be filed covering the plasmids and the cells, with Smith and Jones as co-inventors. The only requirement is that each listed co-inventor must be a co-inventor of at least one claim.

By creative but careful drafting of the claims, it may now be possible to add nearly anyone who was substantially involved in the development of the invention (if that is the goal of the lead scientists). For example, if a technician contributed by optimizing the process parameters in a particular reaction, then a claim can be narrowly drawn to that particular step, using those particular parameters. If the technician is a co-inventor of that particular claim, he or she is a legitimate co-inventor of the entire patent. However, this approach requires the cooperation of the primary inventors, who might not want to see their recognition diluted by adding a long list of co-inventors.

If a patent application claims more than one invention, the patent examiner can require the claims to be split into two or more applications. One group of claims is chosen, and a divisional application directed to the nonelected claims can be filed with a new filing fee. All inventorship decisions should be reviewed if a restriction requirement is received, and they should be revised if necessary to reflect accurately the proper inventorship of each application. If more than two or three inventors are listed on an application, it might be prudent to put a memo in the file explaining what each person contributed. That would make it easier to amend or defend the inventorship decisions, if the need arose. Challenges to inventorship usually do not arise for several years, and it can be surprising to see how differently people remember a series of complex events when personal recognition is at stake.

One point of uncertainty does exist within the law and regulations, however. The regulation adopted by the Patent Office in Title 37 of the Code of Federal Regulations (hereafter referred to as 37 C.F.R.), sec. 1.45 states, "If multiple inventors are named in an application, each named inventor must have made a contribution, individually or jointly, to the subject matter of at least one claim of the application." Some patent attorneys argue that this regulation requires at least one claim to which all the inventors contributed, such as Claim 1. However, I do not agree with that interpretation. If Smith is an inventor of any claim in the application, she satisfies the words of the regulation. And if Jones is an inventor of any claim, he also satisfies the regulation. If the Patent Office wants something else, it should redraft its regulation.

Continuation-in-Part Applications

If an improvement is made on an invention that is already covered by a patent application, the improvement can be claimed by filing a follow-up application called a *continuation-in-part* or *CIP* application. The CIP has a dual filing date; the old information is entitled to the benefit of the first filing date, and the new information receives the new filing date.

A CIP application has one major advantage: the "parent" application is not regarded as "prior art" against the improvement (that is, the information in the parent application is not previously developed technology that might render the improvement unpatentable). The CIP procedure gives inventors incentive to bring improvements to the attention of the public, rather than patenting a basic invention and then keeping subsequent improvements as trade secrets.

But before 1984 CIP applications suffered from a major problem: every inventor of the parent case had to be an inventor of the CIP application. If not, the CIP could not take advantage of the parent case; it would have to be filed on its own, and it would have to be patentable over and above the first invention.

To understand how this created problems, suppose Smith and Jones created a new chemical process and filed a patent application. Then a new researcher, Wilson, joins their lab. With their guidance, Wilson improves the process and increases the yields. Under the old law, the improvement would have to stand on its own; if it was not patentable over the first invention, the three of them would be forced to surrender all rights to the improvement when they publish it, or they would have to conceal it and keep it as a trade secret.

Under the 1984 revision, a CIP application can be filed if at least one inventor on the parent case is also a co-inventor of the CIP (35 U.S.C. sec. 120 and 37 C.F.R. sec. 1.60). This means that Smith, Jones, and Wilson can file a CIP application on the improvement, and the improvement will not have to treat the initial breakthrough by Smith and Jones as prior art. That is the way it should be. They have created an advance in the art; all three of them were involved; and the patent laws should encourage them to tell everyone about it, instead of destroying their rights if they do.

Working in Different Locations at Different Times

The 1984 revision to the patent statute states explicitly that two or more people can be co-inventors even if they do not work together in the same location at the same time (35 U.S.C. sec. 116). Some attorneys deny that there ever was a requirement stipulating that co-inventors must work side by side; however, I have heard some attorneys argue that a consultant should not be considered a co-inventor with a primary inventor, since the two of them did not work "together." The new law removed any uncertainty about that issue and made it easier for outside consultants to be listed as co-inventors with company employees.

What Do These Changes Really Mean?

The loosened rules of inventorship were probably adopted by Congress in response to pressure from the courts. In most lawsuits, the main objective seems to be to do as much damage as possible to your opponent, and patent lawsuits are no exception. Given the driving forces and egos of truly creative people, inventorship is a fertile subject for any attorney who wants to attack a patent. If one pokes around long enough, looking for signs of discontent among people who have not received recognition in a patent, and among people who do not want to share the credit with lesser contributors, one can always unearth a few grubby tubers and seeds awaiting a good rain for a chance to sprout.

The problem is aggravated by the nature of innovation itself. Creativity cannot be measured and forced into cubbyholes without being violated; shades of gray cannot be classified as white or black without creating falsehoods and injustice. It was not surprising, therefore, that in hundreds of patent lawsuits, major battles over inventorship decisions were waged. These battles did not benefit society in any way. Instead, they incurred high costs in legal fees, wasted court time, and often created miserable disputes and bitterness between researchers who were congenial before the aggressive attorney intervened. The fact is that it makes no legal difference how many members of a research team are recognized on the first page of a patent if all of them have already assigned their interests in the patent to a single company or university. Nevertheless, Congress had to step in and say, in effect, "Okay, that's enough nonsense, let's get beyond these petty arguments and get to the real issues."

The underlying theme of the recent amendments seems to be that the courts should not spend their time on issues of inventorship, unless a question of fraud or abuse arises, as long as all of the researchers on a single research team can reach an understanding among themselves. The patent system will in no way benefit from hostile attorneys trying to pry these teams apart, seeking out signs of discontent and opportunities to create unrest.

However, the remaining requirements of the law are not to be ignored. Each co-inventor must still contribute to the conception of at least one claim. A technician who carried out instructions and a manager who provided resources and encouragement will not qualify as inventors unless they have contributed to the conception of the invention, as defined by the claims.

What must be emphasized is that the 1984 changes create new opportunities for all university researchers. The rules have been liberalized, and the changes have cleared a complex obstacle course that once prevented creative people from receiving the recognition they deserved.

4. Ethical Issues in Proprietary Restrictions on Research Results

Alan H. Goldman

The longstanding debate over patents and other proprietary restrictions on the results of university-based research has been rekindled by developments in biotechnology with enormous economic potential (Weiner this volume). The debate has taken on added significance as commercial ties between university scientists and outside corporations have also increased. This chapter assesses the value issues raised by such ties, focusing initially on the option of patents for intellectual products.

Patents make up one possible form of proprietary protection for the results of university research. The alternatives include (1) open revelation with no attempt to control the marketing of products, (2) patenting products and depositing patents with foundations connected with the universities, or (3) depositing patents with government or nonprofit agencies unconnected with the universities. Researchers might also keep their research results secret at the request of either the government or private corporations that have financed the research. If patents are taken, licenses to manufacture and distribute products can then be granted exclusively or nonexclusively, with or without expectation of royalties.

Patents on the results of university research can be viewed as proprietary rights over certain forms of newly discovered knowledge. But knowledge, whether considered as a psychological state of the knower or as a description of what is known, is not normally viewed as property. The usual kinds of property—material goods and assets—are consumable, and their possession is naturally exclusive. The possession of knowledge itself cannot be affected by others sharing in it and cannot be used up. Although possession of knowledge is not exclusive in these ways, knowers do sometimes seek to control its spread by deciding whether, when, and how to communicate it.

Patentable subject matter includes inventions and formulas, in the sense that the patent holder can control the use of the formula in synthetic processes. A patent is not strictly an ownership right to knowledge. In legal terms, the physical realization of an invention is controlled by a patent and not knowledge of how to use or construct the invention. A patent's purpose is not to exclude others from having the knowledge. On the contrary, the patent application process demands public disclosure of information. So patents on knowledge grant the right to control the commercial use or to profit from knowledge. It is a class of machines, and not the knowledge they manifest, that is protected by the patent.

Patents differ from usual ownership rights in other ways as well. For example, they are limited in time. Despite these differences, rights to patents may be justified by the same sorts of considerations of desert and utility that justify other forms of property. In examining the controversial case of patents for products of university research, one needs to evaluate these justifications.

A patent may be viewed as a grant of a limited monopoly to an inventor in exchange for public disclosure of the invention. Once the patent has been awarded, the inventor may then sell his or her rights in the product, or contract with others to manufacture and distribute it. During the life of the patent the inventor can control the product's supply by personally marketing it or by licensing others exclusively or nonexclusively (usually in return for royalties) to market it. By controlling the product's supply, the patent holder or licensee can effectively control price as well. Such control entails the possibility of abusing the public interest, which could prove especially dangerous in the case of products relating to public health. Possibly offsetting this danger is a potential for quality control: a morally responsible patent holder can prevent the market from being flooded with substandard or dangerous products that make use of his innovations. As with any monopoly, however, the likely economic effect of a patent is underutilization of the product relative to demand and a higher price than under free competition.

Despite this effect, there are both economic and noneconomic arguments in favor of a patent system that grants inventors exclusive rights over their products and encourages inventors to carry out the patent application process. The economic arguments, as usual, presuppose an underlying utilitarian moral framework and may be conjoined with other appeals to collective social utility—for example to a need for quality control. All such arguments are *forward-looking* in that they appeal to optimization of total future consequences as justification. In contrast, *backward-looking* arguments are grounded on past events. They invoke just rewards that the inventor deserves for past effort and investment.

An assessment of these arguments for patents is facilitated by studying their origin in contexts less complex than the modern research environment. I propose, therefore, to begin with the case of the single, self-financed inventor working independently. I will then compare that case to groups of researchers salaried by private industry, and finally to university researchers who are partly financed through government grants. I will also note the special case of innovating (partly natural) products related to public health. The moral basis of the arguments for and against patents will emerge from an appreciation of their origin in simple paradigm cases, and of the ways in which the arguments must be modified when applied to the context of university research. In each section, I will discuss backward-looking considerations and then forward-looking considerations.

The Individual Researcher

I will begin with the backward-looking arguments that justify patents as rewards for an inventor's initiative, labor, investment, and accomplishment. These rewards are ideally proportioned to the creation's benefit to society. The basic moral premise of this retrospective claim is a supposed right of a laborer to the fruits of his or her labor, which may perhaps be considered as an extension of the person's activity and therefore of the person. Insofar as this argument merely establishes that laborers should be able to use the knowledge they create, the argument is of little interest to the justification of patents. The creator of knowledge cannot help but possess that knowledge. But the backward-looking argument goes further in the case of commercially valuable knowledge. The further justification of property rights is grounded in a basic intuition that productive effort merits proportionate reward (Becker 1977). The reward for commercially valuable knowledge is the exclusive right to direct profits from use over a specified period of time.

There are several problems with this backward-looking justification, even in the present simple context. First, property justified by value created is problematic when the institution of property results in a mixture of benefits to some and losses to others. (Hence John Locke, the philosopher notable for using backward-looking arguments to justify property, qualified his general support of private property with a proviso for cases when the property rights applied to important and limited resources.)

In the case of patents, the value created includes new products and (since patents require disclosure) knowledge upon which others can expand. On the other hand, there are social disutilities, if the patent preempts work on similar inventions that would have been forthcoming from others and if rivals lose their ability to compete in a market dominated by a patent monopoly. Second, exclusive rights to profits may be excessive, given that the originator's opportunity to be the first to market an invention may in itself constitute an advantage. Although some intuitively object when others ride free on the efforts of the inventor, to permit free riding is to reward productivity, and not to deprive the inventor of reward for the expense or effort of his research. Inventors themselves often depend on prior basic research, although the rewards from patents are not shared with those who provided the necessary background knowledge.

In light of these difficulties, it may be more convincing in backward-looking terms to view patents as payments of contractual debts to inventors. Since patent protection is often an alternative to exploiting a process or product kept as a trade secret, a patent can also be considered as a contractual agreement between society and the inventor. The inventor's contractual obligation is to disclose knowledge (or to relinquish a right to secrecy), and the state's obligation is to provide patent protection. The right to patent, then, depends on a prior right to keep an invention secret, which follows from a more general right to privacy in one's activities. This prior right to secrecy is legally recognized in other contractual arrangements. The courts uphold an inventor's right to provide an invention in confidence to a manufacturer and will award damages if the manufacturer later uses or reveals it without permission. Thus the backward-looking argument appeals either to property rights or to rights of privacy and secrecy, with contractual rights obtained in exchange for disclosure.

These moral and legal rights can be challenged or overridden in contexts in which the products of invention are urgently needed by the public—for health or nutrition, for example. In these contexts the public may have a right to supplies of the needed goods at fair prices, which preempts the right to keep secrets or to own the fruits of one's labor. A right to patent also can be challenged when identical or very similar inventions are brought forth almost simultaneously, an occurrence that is more common than might be imagined. In that case, the patent system grants the entire reward to the first inventor, despite an apparent injustice.

The proper weighting of rights of potential users against those of inventors or suppliers is highly controversial in the context of public needs. According to such libertarians as Robert Nozick, so long as the public does not become worse off than it was before, an inventor can keep her discoveries to herself without violating the rights of others (Nozick 1974). No one has a right to another's time or labor, even if it is desperately needed. Hence no one has a right to the fruits of another's labor. If an inventor chooses to patent and monopolize a product, then the public is better off than it was before (at least until the time when the invention would have been made by someone else). Society is being offered an alternative not previously available. However, more liberal philosophers argue that everyone has a right to have basic needs fulfilled and that this right takes priority over the marginal property rights of others. Thus, if an invention serves such a basic need as health, then its suppression or monopolization under a patent may be disallowed.

I have argued elsewhere that libertarians assume an exaggerated moral distinction between harming and failing to aid (prohibiting the former and allowing the latter). They fail to acknowledge that the fulfillment of basic needs is required for an individual's right to live autonomously (Goldman 1976). One's evaluation of the backward-looking arguments for patent rights in these contexts will depend partly on how one interprets and orders these more basic moral and economic rights.

I now turn to the forward-looking or utilitarian arguments for granting patent rights to independent inventors. The availability of patents is said to stimulate invention by encouraging individuals to invest their time, labor, and money in seeking the rewards made possible by the patent system. Inventions brought forth in this way may also stimulate design of coordinate or alternative products and processes and promote competition with the patentee, despite his monopoly. Furthermore, the disclosure required by patents facilitates improvements, further inventions, technological progress, and competitive research efforts. The public will ultimately receive more goods at lower prices, due to the development of both new products and cheaper production processes through improved technology. Thus, consumers might benefit collectively from the incentives offered by a patent system.

And yet, one may question whether inventors in fact require the incentive of prospective monopoly, and whether their activity is normally motivated by profit. Many inventors may be motivated by the curiosity to probe and the thrill of discovery, along with lesser economic inducement. The question that must be addressed at the margin is whether additional invention will be brought forth by the incentive of patents, and whether this future addition merits limits on supplies for inventions that would be marketed without this inducement. The utilitarian argument balances the disutility of monopolistic control against the utility of accumulation of knowledge resulting from the added motivation to publish or disclose. The argument justifies the legal prohibition of patents when it concerns established technology or inventions that have been developed and withheld for a period of time.

The forward-looking arguments considered so far all suggest that the overall effect of patents is to encourage advancement in technology. This may be more true in infant industries than in those which are highly developed and technologically sophisticated. When numerous patents already exist in a field, the effect may be to stimulate the development of inferior alternatives by competitors, rather than useful invention. Research and further invention may be discouraged also when researchers must scrutinize hundreds of patents to be sure that they are avoiding infringements.

In short, the effect of patents in an industry varies with its stage of development, the nature of its products, the pace of change in the industry, the possibility of trade secrets as an alternative, the ease of inventing around the patents, technological connections among products, and so on. Even in the relatively simple context of the independent inventor, utilitarian arguments become difficult to assess because of the complex and uncertain empirical (economic and psychological) claims underlying them.

In the health sector, as was shown for the backward-looking arguments, the issues are yet more complex. First, patent holders in health-related areas often argue that the patent permits them to control and maintain the quality of the product. However, this assumes that the patent holder is the most motivated and capable party to maintain quality. In fact, the patent holder may wish to hide defects. Second, patenting may shift research from further testing and work on the patented item to the development of alternatives. Third, although patenting may be preferable to secrecy on all these counts, it may not be preferable to open revelation without patenting. Health issues may be so critical that they demand uncompensated disclosure of new techniques. Fourth, although a patent system generally promotes development across an industry, a particular researcher who invents or discovers a new drug or process may not maximize utility by availing himself or herself of the system. Finally, patenting for such purposes as quality control must be compared to all available alternatives, including regulation—competent or otherwise—by government agencies. Since very little health-related research is (or can be) performed by individual researchers, these issues are really more important to the discussions of team research in the private sector and university research.

The Research Team

Group research performed in the laboratories of private corporations differs from that of the independent researcher in regard to both desert and incentive. In the corporate setting, there is often no single inventor responsible for a product or process. Even when there is, the inventor usually assigns all patent rights to the corporation of which he or she is an employee. Many corporations discourage differential rewards to individuals within research teams, because they believe that such rewards encourage secrecy and lessen cooperation among colleagues. In this environment, the patent functions differently: it is no longer a just reward or incentive for individual researchers. Their income from the corporations that employ them (both salaries and bonuses for successful research) may suffice as both just reward and incentive.

Nevertheless, both the backward- and forward-looking arguments in favor of patents for individual inventors have parallels on the corporate level in relation to investment, to property rights, and to returns on investment. A utilitarian can argue that firms will not invest in research if others can later copy the results without the same initial expense. Investment costs must be recoverable, and a fair return for development costs may entail the exclusion of free-riding competitors.

Even the right to secrecy (and therefore the right to patent as a contractual alternative) has a counterpart in the corporate sphere, although based on different grounds. The individual's right to privacy (and the consequent right to keep personal activities secret) cannot justify corporate secrecy. A corporate right to trade secrets derives from its other property rights, including its investment in the resources that are assigned to research and development. Trade secrets themselves are treated as property—they are bought, sold, and sometimes stolen (Bok 1982). Hence in this case the right to secrecy derives entirely from the other backward-looking arguments regarding just returns on investment and adds nothing to them. Nevertheless, a patent may, once again, be considered an alternative to secrecy, one that may not avoid the legal costs of a system based on secrets, but that creates an environment unsuitable to some unsavory practices, such as industrial spying.

What motivates, and to a large extent lends plausibility to, many backward- and forward-looking arguments is the possibility of free riders. Free riders gain the benefits of others' innovations without paying the costs. The issue of free riders is acute because knowledge can be easily appropriated and used in the absence of special proprietary protection (Arrow 1962). At issue is whether knowledge is to be an economic commodity that can be purchased, invested in, and protected against unauthorized use.

The backward-looking argument based on just property rights is more clear-cut in this context than the forward-looking, utilitarian arguments. One clear argument, at least in the private sector, is that if a corporation has paid to develop knowledge in a certain area for the purpose of its profit potential, and if it has succeeded in acquiring the knowledge, then it ought to have some special right over that knowledge and resulting product. A fair price for the commodity need not be set so as to preclude marginal profit for some free-riding competitor, but must assure a reasonable return to the original inventor on investment cost. This justification of the patent system neither assumes loss of opportunity for others nor that the patent system is the only viable means to achieve a fair return. The backward-looking argument in favor of exclusive property rights over knowledge generated may have to be balanced—in health and nutrition cases—against the needs and subsequent rights of the public. And yet in the private sector, some profit on investment seems appropriate.

The forward-looking arguments are difficult to assess because of the wide variety of possible consequences of patents in different areas of marketing and industrial development. As in the case of the individual inventor, it is claimed that patents encourage corporate investment in research and stimulate the competition to produce alternative and improved technology. Moreover, it is claimed that the knowledge provided in disclosure facilitates technological progress. The returns on a patented product can be reinvested in further research. The availability of patents can facilitate the entry of outsiders with new technology into an industry. Yet, in a highly developed industry where many patents already exist, improvements may have to fit existing technology, and inventors may have to go to great lengths to avoid infringement. Those holding multiple patents can use them as weapons with which to threaten expensive lawsuits.

Funds, otherwise available for research or profit, may have to be set aside by both patent holders and new competitors for contests over infringement.

The effect of patents as an incentive or disincentive to research thus depends on the stage of development within an industry. In particular cases, the effect may depend also on whether further testing and research for a product (for example, a new drug) would be socially useful. A patent monopoly, used for quality control and to restrict supply, can be beneficial or detrimental, depending on the motives and abilities of the patent holder and on the availability of alternative means to the positive social benefits. Even if the patent system acts as a general incentive, it need not create benefits in each patent case. Benefits depend also on such matters as licensing policy and how the patent is used.

University Research

I have listed and classified several moral considerations, without fully attempting the difficult task of weighing and balancing them. I now finally turn my attention directly to the complex and controversial context of university research. That this context differs markedly from those previously examined is immediately clear from reconsideration of one of the arguments in favor of patent rights for both independent inventors and corporations. I argued above that rights to secrecy (based on a right to privacy in one context and on trade secrets as property rights in the other) can be a basis for patent rights viewed as a contractual alternative. However, there is no obvious ground for universities to claim that right to secrecy. Indeed, the very idea seems inimical to the whole enterprise of university research. Hence one backward-looking argument that figures prominently above seems entirely out of place in this new context.

However, there are new backward- and forward-looking considerations to add to the previous list. Although universities resemble private businesses in that most research is conducted by teams, the financial resources for university research may come from different sources. Research may be financed by the university itself, by the government, or by the private sector. The particular mixture of sources in given cases may affect the moral arguments.

University employment arrangements also alter both incentives and deserts for work. Researchers are paid and rewarded for achievement with promotions, higher salaries, and recognition within their fields. They are motivated by all these goals, but also by the goal of advancing knowledge itself. In the absence of empirical data, I would guess that very few professors would be moved primarily by the prospect of a patent monopoly over a particular product, especially given the fact that patents are normally assigned to a sponsoring foundation, university, or corporation rather than to the individual professor. University research has for centuries yielded major results without the incentive of patents, and it continues to do so.

Incentives and desert for profit acquire a special significance when the government sponsors research under grant programs. Publicly financed work must seemingly give priority to the public interest in the disposition of its results. Incentive is provided by the grant itself and by the prospect of its continuation or of others, not by the prospect of future returns. It is the public that deserves the return from its own investment in this case. This is doubly true when the research not only is publicly financed but also results in products related to public-health needs.

Consideration of university research financed by private industry returns one to a framework in which the principal backward-looking argument appeals to a fair return on investment. But the question now arises whether such direct financing should be encouraged or accepted by a university. The answer in particular cases will depend on the specificity of the grants or the directness of the ties between the grant and the products and profits of the funding corporations. Some cases raise the related issue of whether universities should establish private corporations or foundations to manage patents and licenses for marketing the products of their research.

Forward-looking arguments for university foundations focus on the need for additional funding for research, especially in times when government support dwindles, and on the possibility that profitable research can create funds to support research with little profit potential. The backward-looking argument points out that university professors remain underpaid relative to other professionals and deserve additional resources form the private sector, especially when their research happens to have profit potential. On their side, commercial firms welcome university participation as a way of facilitating commercial development. Any full assessment of the options must also consider the legal and administrative costs of patent management.

The main debate continues to center on whether commercial interests are compatible with a special professional ethic, developed over centuries and rooted in deep moral commitments, that has become the creed of academic administrators and researchers. Of special concern is the effect of such interests in slanting research in particular directions and suppressing other research. At issue is the relevance of the sorts of products likely to be developed in university laboratories to academic as against commercial interests. The special ethic to which I refer places the advancement of knowledge above other interests—political or commercial—and endorses open, disinterested, cooperative inquiry both as a means to that goal and as a goal in itself. In contrast, utilitarians view the advancement of knowledge as a means to collective happiness and therefore welcome the transition from pure to applied research, when this does not result in fewer applications or practical benefits in the future.

I believe, however, that utilitarianism ignores and distorts the working ethic of scientists and other inquirers, who tend to pursue knowledge not simply as a means, but as a separate end in itself. Indeed, one common criticism of (hedonistic) utilitarianism notes its inability to account either for ends other than happiness or for the depth of alternate moral commitments. Many working researchers are willing to forgo personal pleasures, self-interest, and cumulative happiness in pursuit of the knowledge they seek. And few think of that pursuit as aimed at or even limited by the collective happiness of society.

Given the depth of this commitment, it is no wonder that many adherents of the ethic view with alarm the possibility that commercial or political interests will intrude into academia and divert their colleagues from the open pursuit of knowledge. They see the openness of inquiry—the free sharing of information and theories—as an end in itself, a model of an open and free society. Open communication and dialogue are also a means in that they not only add efficiently to the existing body of knowledge, but also allow for prompt criticism and early correction of errors. One may wonder whether any of these considerations is genuinely ethical or moral; given the independent value of knowledge and the depth of commitment to its pursuit by those in the academic profession, I would answer that question affirmatively.

In reply, it is argued that this model of scientific and humanistic inquiry is little more than romantic idealization. In the real academic world there are competition, prepublication secrecy, and political cliques in academic power structures (journal editors, department chairmen, grant administrators, and so on)—all of which point to political as well as financial motives underlying many research programs. Thomas Kuhn and the sociologists of science who followed him showed the degree to which normal research is governed by ruling paradigms or models, and how these entrenched paradigms are maintained by the inner circles of the disciplines they govern (Kuhn 1970). However, one can follow Kuhn in abandoning the notion of totally disinterested science while still maintaining a distinction between the development of a discipline from within (based on its own central values and concerns) and the commercialization of a discipline from without (in pursuit of marketable products). The latter may be aptly compared to political influence in the purer sense of politics, including, for example, politically inspired censorship of publications.

Several further replies can be considered here. First, there may be room for both pure and applied research within universities. Both are conducted in technical and agricultural universities, even in the absence of commercial interests. Second, the line between pure and applied research is vague and the transition from one to the other much quicker in present disciplines—such as biochemistry—than it was in their predecessor disciplines. In light of this situation, scientists who simply ignore the commercial potential of their work might be seen as naive or even socially irresponsible. Third, whereas political censorship is coercive and distorts the truth, commercial arrangements between professors or administrators and private corporations are voluntary. Applied research need not, and if successful probably will not, distort the truth. Fourth, commercial connections can contribute funds to pure research and even to research in nonscientific disciplines without corrupting the ongoing overall pattern of research within the university. This is possible when the connections are suitably indirect.

Finally, let me evaluate these objections to commercial ties and the replies to them. First, there seems to be little doubt that commercial ties between private corporations or foundations and universities can affect the openness of the research process. Where corporations pay researchers directly for specific research, trade secrets may be required. Even when patents function as an alternative to the prolonged maintenance of corporate trade secrets, secrecy is demanded during the process of developing patentable products and seeking the patents. And despite the requirement of disclosure for patents, inventors sometimes disclose less than fully. Although researchers in the absence of commercial influences may hide their results from rivals until publication, the intrusion of profit motives, whether directed toward patents or trade secrets, can only further deteriorate the ideal of a fully open, cooperative pursuit of knowledge.

Second, while there may be room for both pure and applied research within universities and while the line between the two is often blurred, it seems clear that the profit motive, even when exerted indirectly through a foundation, can slant the direction of research toward areas with a potential for profit. This bias benefits financially departments with applied bents and disciplines pursuing pure research likely to lead to marketable applications. It strengthens departmental factions whose research is at the applied end of the spectrum. Within the applied sciences, research may be directed away from domains where patents already exist. Greater financial inducements may also lure administrators away from the university proper or take administrative time and energy away from broader university concerns. Utilitarians, as I have shown, argue that the main benefit of research lies in its applications, and so may consider this a healthy shift of emphasis. Although utilitarians can qualify their endorsement of applied research in recognition of the possiblity that basic research can have major long-range social benefits, others just dismiss the utilitarian view of the advance of knowledge with its emphasis on application.

Third, although connections between universities and patent-holding foundations need neither influence the direction of research within any specific field nor slant attention and resources away from other research fields, commercial interests constitute at least a temptation in those directions. A foundation connected to a university may begin with the purest of intentions—for example, to maintain quality control, prevent fraud in the marketing of products, and aid its university with financial support. It may wisely seek to accomplish these aims by holding patents and granting nonexclusive licenses to all who maintain quality. But the benefits to the university may accrue mainly in those areas whose research is relevant to the financial interests of the foundation. And the foundation may then directly or indirectly suppress research inimical to its interests—a cardinal sin from the perspective of the academic ethic.

In this regard, consider the connection between the University of Wisconsin and its foundation, WARF (Weiner this volume). In 1944 a lawsuit arose from the refusal of WARF to license oleomargarine manufacturers for their irradiation process. A U.S. court of appeals ruled that WARF improperly used its patents against the public interest by depriving consumers of access to vitamin-enriched margarine. The history of this financially successful research foundation should give pause to those who see no inherent conflict between the mission of a university and the marketing of research results by a foundation operating in its name and on its behalf. Controversies continue to cloud such connections. A relevant instance is MIT's association with the Whitehead Institute (see Boonin this volume).

Overview

I shall now summarize the major considerations for and against patents (by now perhaps too numerous to keep track of) in the context of university research. The backward-looking arguments that justify patent rights for independent inventors and private corporations have only a tenuous application in the university context. These arguments appeal to deeper rights to (1) the fruits of one's labor, (2) privacy or secrecy, and (3) fair returns on successful, profit-motivated investment.

However, appeal to those rights seems out of place in the context of university research and its fruits. Professors, who are employed to conduct research, do not appear to have exclusive rights to profits from patents on marketable products (beyond profits from the publication of the ideas themselves). If research is government supported, then neither the researchers nor their universities appear to have moral rights to monopolistic control over resulting products. The same conclusion holds if the universities are state supported. If private universities finance research internally, then they, together with their professors, may have rights to patent, and may share the returns in some fair proportions. But whether they ought to exercise these rights depends, in part, on forward-looking considerations, on the nature of the products, and on possible rights of public access to the products. For example, liberal social philosophers recognize a basic, public right to products needed for medical purposes at the lowest possible prices. These rights may imply that universities would be wrong to exercise legal rights to patent in such cases, especially if quality control can be achieved in other ways.

The forward-looking arguments are complex and indecisive, even in the context of private business. The main additional factor I considered in the university context was the influence of royalty-seeking, patent-holding foundations upon the direction, content, and process of research. Extraneous commercial interests can tempt researchers away from the ideal of disinterested and fully open inquiry. Although there is no objective moral imperative commanding maintenance or even pursuit of that ideal, professional academicians have been committed for centuries to open inquiry, and that commitment will certainly be felt as moral by them. In the university context, it has a profound ethical significance. Just as those committed to this ideal ought to be wary of the intrusion of external, political influences, and ought to resist tooth and nail the suppression of research inimical to particular political factions, so they ought to be wary of the intrusion of direct commercial interests. History shows that these can have similar effects.

Entrepreneurs will view this resistance as a naive, anachronistic attempt to reinstate an already lost, romanticized ideal of the university as an ivory tower. But the ideal in question has survived many sustained and direct attacks over the centuries and produced cultural offspring that may be more significant in the long run than the recent transformations of culture by the entrepreneurial class.

Nevertheless, avoidance at all costs of the commercial ties considered here *would* be naive and ideological (in the pejorative sense); the value of disinterested inquiry, as free as possible from distorting extraneous influences, must be balanced against other forward-looking considerations. For some financially strapped private universities, one such consideration may be survival itself. That situation is common although, for better or worse, it is not one that often gives birth to highly profitable research. Other considerations relating to the health and biochemical industries—especially in genetic technology—make it clear that forward-looking arguments will continue to be difficult to assess, given the number of complex and controversial factors that must be estimated and weighed.

References

Arrow, K. J. 1962. "Economic Welfare and the Allocation of Resources for Invention." In *The Rate and Direction of Inventive Activity*, edited by R. R. Nelson, pp. 609–624. Princeton: Princeton University Press.

Becker, L. 1977. *Property Rights: Philosophical Foundations*. London: Routledge and Kegan Paul, chap. 4.

Bok, S. 1982. *Secrets*. New York: Pantheon Books, chap. 10.

Goldman, A. 1976. "The Entitlement Theory of Distributive Justice." *Journal of Philosophy* 73: 823–835.
Kuhn, T. 1970. *The Structure of Scientific Revolutions*. Chicago: University of Chicago Press.
Nozick, R. 1974. *Anarchy, State, and Utopia*. New York: Basic Books, chap. 7.

Historical Case Studies

PART II

While assessing current controversies about intellectual property, it is important to understand the history of efforts to protect innovations. Many of the current issues about stimulating technological research and encouraging the implementation of new discoveries arose in decades past. It will help to know which expectations were confounded and what pitfalls were encountered as attempts were made to adjust the property system to changing science. Medicine and agriculture are particularly interesting in that innovations in those areas gave rise to debates focusing on noncommercial research settings and goods not previously covered by intellectual property law. A history of controversies over intellectual property and of the evolution of our present system from such controversies sets the context for discussion of current problems. In dealing with new challenges to the intellectual property system, we should take advantage of hindsight available from earlier debates about the basis for protection and the likely consequences of adjusting or expanding the scope of intellectual property protection.

The two historical papers begin with events early in the twentieth century. Charles Weiner analyzes several episodes in the history of biomedical research that engendered passionate disputes over the appropriateness of exclusive rights to medically important innovations, particularly when the rights are held by universities and similar public institutions with an altruistic commitment to serve society. One side argues that proprietary interest poses a threat to openness, skews university research toward the marketplace, and generates conflicts of interest for academics. The other side stresses that the use of patents facilitates technology transfer from the university to society at large, assures quality control over medical products, and generates new sources of revenue to support fundamental research. The issues are particularly acute for biomedical scientists who see themselves as committed to research that meets sensitive social needs and claim that their results are of special human value.

Weiner brings out distinctive differences among the cases he considers. He contrasts an incident in which the option to patent was chosen with one in which it was, in the end, rejected. The first incident involved research physicians who elected to patent a scarlet fever antitoxin to maintain quality control. They became embroiled in costly legal battles to protect the patent. Eventually, respected figures in the international health community concluded that the effect of the patent had been to discourage and prevent research. The second episode concerned a discovery at Harvard University of a medically useful liver extract. Faced with a forced choice between patenting or putting the discovery in the public domain, the university finally decided not to patent. It later appeared, however, that the absence of a basic patent as a basis for industrial patents had hindered manufacture and posed an obstacle to standardization and quality control. In both instances, expectations about the benefits of patent protection were confounded and both the decision to patent and the decision not to patent had unwelcome consequences for the course of research and development.

Weiner's examination of the activities of an independent foundation established by the University of Wisconsin for patent management reveals an ongoing series of both beneficial and detrimental outcomes. The foundation is still in existence and its overall benefit is still disputed. His case studies indicate that, in the complexities of real noncommercial research situations, general claims and presumptions have only limited usefulness for deciding whether to patent. The cases suggest that specific goals should be identified and pursued directly, with proprietary protection in view as only one of a number of options for achieving the specific goals.

Weiner concludes his historical study with an analysis of a far-reaching review of the patent system commissioned from a university researcher for a congressional committee on patents, trademarks, and copyrights. The review concluded that the system was a hindrance rather than an impetus to research. A rebuttal on behalf of an association of patent lawyers argued that the review concentrated on large companies and omitted smaller organizations where patenting was important. The rebuttal also suggested that without a patent system, there would be more secrecy, not more openness. It is notoriously difficult to get information from companies about the benefits they gain from patents and also daunting to devise tests for the claim that without patents there would be less openness. This instance highlights the obstacles to bringing forth evidence to settle the issues in dispute.

In their narrative, Frederick Buttel and Jill Belsky cover the long struggle by the agricultural seed industry to gain the proprietary protection needed to make plants a profitable commodity. As long as plants are propagated naturally and plant offspring tend to resemble their parents genetically, it is difficult to establish proprietary control over plant varieties. Given this natural barrier to private profit and private investment in the development of new plant varieties, the public research system, including land-grant universities, began early in the twentieth century to play an important role in plant breeding and in releasing new varieties. Using scientific breeding practices, the public research system had successes that stimulated private-sector interest in breeding. Hybridization—its most dramatic success—drastically altered the market environment. The nature of hybrids, which forces farmers to return to the market each year to buy seeds, made possible proprietary control.

Ironically, the general success of the public breeding program led to the revitalization of the private seed industry, which then began to pressure public breeders to stop releasing new varieties. The rejuvenated industry's lobbying efforts led to the creation of sui generis patentlike protections enacted for the seed industry. These protections were later extended and, the authors contend, eventually helped promote incorporation of seed companies as subsidiaries of large multinational agrochemical companies. But the protections were less effective in stimulating research, breeding, and release of genuinely new varieties, according to Buttel and Belsky.

If the authors' conclusions are correct, this is a story of research success in a noncommercial setting bringing vitality to the private sector, which then successfully lobbies for new proprietary protection and the constriction of public research. Whether this is a negative outcome, as the authors contend, is a disputed issue. However, the story does exhibit a greater complexity and subtlety in the interconnections between noncommercial research and the commercial development of that research than usually envisaged.

5. Patenting and Academic Research: Historical Case Studies

Charles Weiner

In the early part of the twentieth century, academic researchers rarely patented their results. There was widespread fear that the effects of patenting would commercialize the academic environment, corrupt the traditional nature of the university, and impinge on the public's trust and support of university researchers. By the 1920s, however, hard-line opposition to patenting among academic scientists was beginning to be modified in favor of a more flexible approach. It was no longer sufficient to harp on the oft-recounted ill effects of patenting: altering allegiances and aims, skewing research, dampening the incentive for information exchange among colleagues, generating professional and financial jealousies, or focusing creative energies on profit rather than research.

Proponents argued that there were other factors to be considered. In their view, patenting was an efficient way to translate knowledge and processes into useful products. It provided the means not just to exercise a financial monopoly but also to control the quality of products and so protect the public from exploitation. They argued that patents could generate new sources of revenue to strengthen academic research and could provide scientists with an incentive for further efforts through financial reward and public recognition.

The situation posed a very real dilemma for academic scientists, and nowhere more clearly than in the field of biomedicine. The controversy involved the real and perceived tensions between the tradition of making medical discoveries freely available for public use and further development, and the emphasis in the patenting system on monopoly, control, and the income-producing potential of inventions. Arguments went back and forth: Shouldn't means other than patenting—for example, prompt publication of results and diligence against potential research monopolies—be followed in order to maintain the field's special status and commitment to developing knowledge for public health and medicine? Shouldn't changing circumstances—the need for innovative sources of academic research funding, the increasing relevance of academic biomedical research to successful public health applications, the demand for added control over production methods to assure quality and prevent exploitation, the increasing time and expense needed to bring an innovation into public use—require a reassessment of patenting as a potential tool in the public interest?

The debate raged on—in a particularly intense way from the turn of the century until just after World War II.[1] Some compromise was inevitable; approaches and attitudes were modified over time from a virtual taboo on patenting; later, while it never became acceptable to some academic biomedical researchers or a source of significant concern to others, patenting was carried out with reservations in the context of stringent policies and procedures.

Concern over the issue of patenting academic research has been most evident during periods of economic crisis, such as the depression of the 1930s, and at times when significant new research made possible applications of major significance to society (as with various vitamins, vaccines, and hormones developed during the 1920s and 1930s, and in the 1970s with the emergence of the new biotechnology). During such periods, when assumptions, traditions, and values are being reassessed, restated, or restructured, the participants often look to the past for help in understanding their own situations. Historical precedents or problems are cited, sometimes as support for either a defense or a revision of existing policy. But history has been used in this way mainly to grind axes rather than to explore and understand the situation in all its complexity.

The following accounts of several historical cases are based on archival and published records. They illuminate the continuing controversy over patenting and past experiments in academic patent management. They emphasize the motivations, expectations, and results for scientists, their institutions, and the public. These historical accounts include an early innovative but unsuccessful patenting plan at the University of California, the highly controversial scarlet fever antitoxin patent, Harvard University's decision not to patent a pernicious anemia treatment in the 1920s, the consequences of the financially successful "Wisconsin plan," and Seymour Melman's provocative 1950s study of the impact of patents on research.[2]

Unfulfilled Expectations

In 1905 T. Brailsford Robertson, a young graduate of the University of Adelaide in Australia, took a junior staff appointment in Jacques Loeb's department at the University of California–Berkeley.[3] Within two years he earned a Ph.D. and was promoted to the faculty as assistant professor, but he quickly grew dissatisfied with the heavy teaching load and lack of support for research at the university. In reaction, he threw himself headlong into a variety of research projects, including the development of a fermentation process that he considered patenting in 1906. A friendship with the chemist Frederick G. Cottrell grew during the next few years, and one of their recurrent points of discussion was the possibility of creating a patent scheme in which the royalties would be used to fund research laboratories and projects. Cottrell, also a faculty member at Berkeley, took out in 1907 a patent on an electrical precipitation process for cleaning smokestack emissions. According to Cottrell's account, he offered the rights to the university, which refused them on the grounds that the university charter did not permit involvement in commercial ventures. Cottrell finally decided to set up an independent nonprofit group. Research Corporation, established in 1912, managed scientific patents and used the surplus income to support further research. Robertson's own feeling at the time was that a patent scheme might best be developed in an industrial or commercial setting, rather than at a university.[4]

Following Loeb's departure in 1910 for the Rockefeller Institute of Medical Research, Robertson succeeded to the chair of biochemistry at Berkeley, and in 1915 became the head of the Department of Biochemistry and Pharmacology. He remained dissatisfied with the university, however, and devoted a considerable amount of time to a new patent project—the development of a chemical substance useful in controlling the growth of prickly pear, a plant troublesome to farmers and ranchers in Australia. Loeb tried to discourage Robertson's commercial leanings by promising to help him seek out a more suitable academic appointment. But Robertson continued to consider taking a job in a private enterprise, for example, at the Mayo Clinic, which he anticipated would provide a more supportive and certainly a more lucrative base for his work. When the Berkeley administration sensed that his intentions were serious, department funding and administrative support increased significantly. Robertson decided to stay.[5]

Around the same time, Robertson's experiments with an extract from the anterior lobe of the pituitary gland in the brain led to the isolation of tethelin, which he believed to be a growth-promoting substance with potential medical applications (Robertson 1916a, b). He applied for a patent in October 1915, and it was granted in March 1917. Clinical experiments were undertaken. The results appeared so promising that the Berkeley news office sent out press releases touting the substance as the possible key to growth enhancement and a cure for cancer (*New York Times* 1917; *Daily Californian* 1917). Although upset by these premature speculations, Robertson had high hopes for tethelin and continued to press its potential medical usefulness.

The idea also occurred to him that the patent income on tethelin could be used to fund further biomedical research. Robertson proposed to donate his patent rights to the university. Profits resulting from commercial licensing of the patent were to be placed in an endowment fund to establish a medical research institute on campus for Robertson and other scientists.[6] The Berkeley regents had reservations at first, based in part on the possible impropriety of a state-funded institution entering into a contract with a private firm. Robertson explored Cottrell's group, Research Corporation, as an alternative in the event that the Berkeley negotiations fell through.[7] The regents finally decided in favor of Robertson, and an agreement was signed in September 1917. Shortly thereafter, the university granted a five-year exclusive license for manufacture and sale of tethelin to the H. K. Mulford Company in Pennsylvania (Robertson 1917).

Tethelin was neither a medical nor a financial success, and no royalty income accrued sufficient to fund research.[8] Robertson grew dissatisfied again with what he considered to be the Berkeley administration's disinterest and lack of support for the work in his department. Loeb resumed efforts to find Robertson another job. This task proved to be more difficult than before, as Robertson's decision to patent tethelin had caused his integrity to be questioned in the academic community. University scientists, Loeb emphasized, were imbued with a "strong prejudice" against patenting—a prejudice based on suspicion of mercenary motives, fear of commercialization of the research process, and concern about the possibility that "pure science will be doomed" if the academic environment were to become tainted by patent consciousness.[9]

Robertson was finally appointed to the biochemistry chair at the University of Toronto, but only after an unusually careful review process in which his decision to patent was labeled "improper" by at least one scientist.[10] Those in favor of the appointment tended to give him the benefit of the doubt in regard to motive, but suggested that he was "misguided" and "injudicious" not to have paid more attention to scientists' longstanding apprehensions over the ethics of patenting. Uppermost in their minds was the "specialness" of biomedical discoveries in relation to public health; they believed that such discoveries should be viewed in quite a different light than, say, inventions in engineering or chemistry.

Robertson stayed at Toronto just one year. The harsh climate, high cost of living, and quality of life in an industrial center did not suit him or his family. In 1919 he returned to Australia to take over the chair at the University of Adelaide, left vacant on the death of his father-in-law, Sir Edward Stirling. Up until his death in 1930, Robertson continued to explore novel research-funding schemes, particularly in connection with biomedicine and agro-industry.[11]

Roberston's hopes to use the tethelin patent to generate income to support his research were unfulfilled, as were the public's expectations that the substance would have important medical applications. His efforts brought criticism from other scientists who believed that he was departing from the existing norms of ethical professional behavior. A few years later, in the mid-1920s, two medical researchers discovered and developed a highly successful therapeutic approach to a major disease, scarlet fever. Their decision to patent, although based on different motivations, made them the subject of similar intense controversy in the international medical community.

The Scarlet Fever Case

George Dick and Gladys Henry, research physicians at the University of Chicago, met in 1911 and married three years later.[12] During World War I, they began a collaborative research program on the etiology of scarlet fever, under the auspices of the McCormick Institute in Chicago. In 1921 they isolated and identified the streptococci that cause the disease. Then, using human volunteers, they injected streptococcal toxin into the bloodstream to induce scarlet fever symptoms. They proceeded by experimenting with different concentrations of the toxin, showing that high dilutions could be used as a skin test to assess susceptibility and that more concentrated solutions could be used in preventive immunization. A major finding in 1924 was that the blood serum of people immunized with the toxin and of patients convalescent from the disease contained an antitoxin that neutralized the toxin. The Dicks determined next that an efficient way to mass-produce the antitoxin was by immunizing horses with the toxin (Dick and Dick 1924a, b).

The Dicks' results aroused considerable interest in the therapeutic potential of their discovery. Manufacturers began flooding the market with assorted antitoxins based on publications by the Dicks, who in turn grew concerned that many of these products were inferior in quality, perhaps dangerous. Officials at the Public Health Service were contacted for advice about how to establish quality control over production. The Public Health Service suggested patenting as the most effective means. The standard ethical barriers to patenting, it noted, had been breaking down as a result of the general acclamation surrounding the insulin arrangement at the University of Toronto, where the discoverers of insulin had taken out a patent and made arrangements to monitor production through university laboratories.[13]

Patent applications on scarlet fever preparations were filed within three weeks of the communication from the PHS. The Dicks intended to follow the Toronto plan by taking out patents in their names and then donating them to a university or similar organization. After considering the various options, the Dicks offered the patents to the American Medical Association. The AMA declined, however, to undertake the responsibility. They had abandoned any thought of serving as a patent-management body following the failure in 1918 of an effort along these lines with thyroxin, the hormonal substance discovered by E. C. Kendall of the Mayo Clinic. The failure had stemmed in part from dissension within the ranks of the AMA membership, which was sharply divided in its opinion on the wisdom of medical patenting. One side suggested that patent income provides the impetus and wherewithal for further discovery; the other side objected in principle to the notion of profit on medical discovery, whatever the reason. Unwilling to revive the specter of this controversy, the AMA advised the Dicks to establish an independent nonprofit committee to hold and use the patents primarily as a way of exercising quality control over production (American Medical Association 1937).[14]

The Scarlet Fever Committee was established early in 1925 with a membership spearheaded by the Dicks themselves. Members of the committee were to receive no compensation, and the royalties were to be used strictly to implement a program of quality control through testing at the institute. Before long, four companies were licensed to manufacture the antitoxin. Everything appeared to be going smoothly. The Dicks were nominated for the Nobel Prize in 1925, but for some reason the Nobel committee decided in the end not to award a prize in medicine and physiology that year.

The medical community was beginning to seethe with criticism regarding the Dicks' decision to patent and thus to exercise control over a substance with significant public health consequences. By 1926 editorials and correspondence in major medical and public health journals were expressing decided objections to the Dicks' actions, with one commentator suggesting that "the whole matter leaves a bad taste in one's mouth" (*American Journal of Public Health* 1926).[15] The primary contention, granted the validity of the Dicks' concern for quality control, was that such control can be exercised as effectively (if not more so) in state and federal laboratories and regulatory agencies. The fiercest outcry of all occurred in Great Britain, where the Dicks had also applied for patents. Articles in *Lancet* and the *British Medical Journal* insisted that the evils of monopoly—restrictions of research and manufacture, high prices, and commercialization of the research process in order to satisfy payoff expectations—were especially dangerous with regard to therapeutic discoveries (*British Medical Journal* 1927a, b, c; *Lancet* 1927c, b).

Despite the ill feeling in the medical community, the Dicks were aggressive in their determination to defend the patents (Dick and Dick 1925, 1927). In 1928 they brought legal action for infringement against the Lederle Antitoxin Laboratories. The case lasted two years and ended with a decision favorable to the Dicks.[16] But this decision only added fuel to the fire. The Dicks were publicly (though in veiled terms) accused by Alan Gregg, director of the medical sciences division of the Rockefeller Foundation and a particularly vocal critic of the trend toward medical patenting, of squandering from patent royalties close to a six-figure sum in legal fees (Gregg 1933). The researchers threatened to sue Gregg unless retractions of such purportedly incriminating statements were published in a prominent manner.[17] At the same time, their vigilant campaign of patent defense continued; for example, in 1933 they threatened legal action against the New York City Health Department and the New York State Laboratories in Albany.[18] Meanwhile, the British medical community continued to back the Dicks' opponents. Testimony came from Sir Walter Fletcher of the British Medical Research Council and from others that the effect of the patent in Britain had been to discourage and prevent research rather than to protect quality.[19] Researchers were reluctant to work on the subject because they feared their research might lead them to impinge inadvertently on patent-protected areas. The League of Nations health organization issued a similar statement.

All this criticism perturbed the Dicks, but it also prompted them to more aggressive tactics and shriller protestations of their "rightness" in the matter. One of their final public defenses took place at a special meeting on patents sponsored by the American Medical Association in 1939 (Dick 1939). Shortly thereafter, the issue was closed for all practical purposes when antibiotics took over as the preferred means of treating streptococcal infections. An antitrust investigation brought against the Dicks on the urging of Alan Gregg was dropped when it became clear that scarlet fever antitoxin was no longer a therapeutic priority.[20]

Gladys and George Dick originally had justified their patenting of the scarlet fever antitoxin because it would enable them to protect the public by maintaining quality control. The response of a significant segment of the medical community was that the public was being harmed because the patent hindered further research in the field. The controversy undoubtedly affected the decisions of other researchers faced with potentially marketable discoveries. Only a few years after the Dicks applied for their patent, a different path was taken at Harvard University. Harvard's decision was similarly motivated by a desire to protect the public, but the approach in this case was not to patent. It, too, led to mixed results.

Harvard's Decision in the 1920s

In 1926 George Minot and William Murphy, faculty members at the Harvard Medical School, demonstrated that liver was an effective treatment for pernicious anemia. The following year, a method of extracting the active ingredients was developed by Minot and Edwin J. Cohn in the medical school's Department of Physicial Chemistry.[21] The Corporation of Harvard College immediately considered alternative strategies of developing the discovery, both to protect the interest of the public and to ensure availability under the best possible conditions of cost and quality. The discussion centered primarily on whether to patent and, if so, what policy to follow. On 12 March 1927 the corporation voted that with Minot and Cohn's consent the university should apply for a patent, license commercial manufacturers on a nonexclusive and royalty-free basis, simultaneously make all information available to anyone for research or other scientific purposes, and work out a way to deposit the patent as soon as possible with a central government body prepared to assume administration of the rights and protection of the public without profit (Cohn 1951)

The action was prompted by President Abbot Lawrence Lowell, who appeared convinced that patenting was the correct route to take. Lowell was advised to proceed cautiously, however, because of the longstanding ethical tradition against patenting in the medical profession and similar doubts among elements of the academic community. He decided that the most sensible course of action was to go ahead but to do so "in consultation with other universities and laboratories so that it should not be an isolated, but a cooperative action."[22] For their part, the inventors maintained a cautious neutrality, pointing out that the matter was one of university policy and should be decided by the Corporation. Their only concern was to ascertain, in time for the next scientific meetings, what that policy would be, so that they would be in a position to state Harvard's intentions "precisely" and avoid any "semblance of mystery or lack of frankness."[23]

By examining precedents, they carefully reappraised the patenting decision. The case of tethelin at Berkeley appeared to be a promising experiment in patent management, but since the discovery had never been used widely and had in fact virtually petered out of clinical usage, the Berkeley experience evidently had little to offer in the way of real insight into the potential problems. On the other hand, the patenting of adrenalin in the 1890s, with its widespread importance and vast legal complications, stood as an example to all that patenting might profit "neither learned institutions nor the public." In the case of insulin at the University of Toronto, control of the process of production was critical, but it remained unclear that a patent was necessary to exercise this control—especially in light of the fact that semigovernmental laboratories were available for the work.[24]

The overall evidence suggested that patenting was not essential in cases where the motive involved concern for the public interest rather than financial gain. Frederick G. Cottrell (who had created the Research Corporation to manage patents) reiterated this point to a faculty member at the University of Michigan, who in turn passed the information on to Harvard officials. Cottrell said that the difficulties of patenting were more trouble than they were worth, at least for such public-oriented institutions as universities and government laboratories. He confessed to having come more and more around to the view that "dedication of patents to the public under the law of 1883 may have a wider field of legitimate usefulness than we were at one time inclined to think," and that with the exception of a few of the most fundamental discoveries, patenting should be forgone altogether and the information simply "published as promptly and fully as possible."[25] This approach would eliminate the need for defensive litigation and protracted control responsibilities, and would mitigate one of the worst effects of patenting—the high prices resulting from the limited competition inherent in the patent-license system.

On the basis of these considerations, and taking into account the adverse publicity that then surrounded the scarlet fever patents, the Harvard Corporation reversed its original vote. Harvard's decision not to apply for a patent on the Minot and Cohn liver extracts was praised by many members of the medical and academic professions. Sir Henry Dale commented on the ease with which British laboratories were undertaking research and production, without the need to contend with the issue of monopoly and its many legal and ethical ramifications.[26] But by 1951 Cohn wrote that subsequent developments had convinced him that the public interest would have been better protected by means of a patent at the outset. He felt that "a simpler relation could have been established" if Harvard had had patents on which industrial patents could be based (Cohn 1951). Furthermore, standardization and control proved harder to achieve without the authority to license qualified manufacturers. The original investigators engaged in extensive correspondence with firms to try to ensure quality, but the exercise often proved frustrating because of distances, communication barriers, and differing motivations. It was a diffuse, at times haphazard, arrangement.[27] A Committee on Pernicious Anemia was formed at Harvard in 1927, but its role was strictly advisory; it had no power to enforce policy.

One event, not mentioned by Cohn in his history, must have had an influence on his reevaluation of Harvard's decision not to patent. In January 1935, Cohn and Minot first learned that Eli Lilly and Company had been granted a patent for a process for making a liver extract for use in the treatment of pernicious anemia. The patent in its essential features followed the Cohn method, which was made known to Lilly's scientists by Minot in 1927 when he was invited to visit the company to help treat the wife of the president of the firm, who suffered from the disease. Lilly applied for its patent in 1928.

When Harvard's Committee on Pernicious Anemia learned of this in 1935, it complained to Harvard's president that the representatives of the company had not been straightforward in their relations with the committee. Lilly had urged Harvard to patent the process, which the university refused to do. Then a Lilly scientist had applied for a patent, and Harvard was not informed. In addition, during the period of the existence of the committee—from 1927 to 1929—the company had marketed the extract and advertised it, not as a product of their own methods, but as being prepared under the direction of the Committee on Pernicious Anemia of the Harvard Medical School. The 1935 revelation of these events was of special interest because Harvard had just established a definitive policy against the patenting of therapeutic measures by its staff, and had promised legal aid to prevent others from doing so.[28]

WARF and the Price of Success

Unlike the Harvard case, a parallel case at the University of Wisconsin led to an aggressive patent ownership and management arrangement, with income for research as a basic motivation. Although launched on the basis of a specific patent, this solution was applied to the development of patents for the university as a whole. And the resulting financial success intensified controversy about the appropriate role of the university and its relation to industry and the public.

In the early 1920s Harry Steenbock, a faculty member in the the College of Agriculture at the University of Wisconsin, developed a way to manipulate and concentrate vitamin A as a dietary supplement.[29] He also discovered that the sun's ultraviolet rays produced the antirachitic vitamin D (Steenbock 1924a, 1927). These discoveries had the potential to enhance the nutritional value of oleomargarine and infant formula, making these products more competitive with natural dairy products. Steenbock was concerned because the Wisconsin dairy industry was a strong supporter of the university and had considerable influence in the state legislature, which controlled the university's budget. He filed patent applications in the summer of 1924.

At the same time, commercial firms began to express an interest in the discovery. Quaker Oats, for example, offered Steenbock close to a million dollars for outright purchase of the rights. Such offers suggested to Steenbock the need for a nonprofit institution to manage the patents. But the University of Wisconsin clung firmly to its traditional hands-off attitude. Steenbock explored other alternatives that were allied to the university but operationally independent of it. He gathered information about institutions where patent management plans had been put into effect and were administered by independent committees loosely attached to the university.[30] Encouraged by what he heard, he urged and finally persuaded H. L. Russell, dean of the College of Argiculture, and Charles Slichter, dean of the Graduate School, to lend their support to the implementation of such a plan at Wisconsin.

Steenbock had argued that the patents could be controlled to protect the dairy interests, and that royalties from licensing could provide substantial support for research. In his public announcement of the vitamin D irradiation process, he stated that the patent would be handled through the university "to protect the interest of the public in the possible commercial use of these findings" (Steenbock 1924b).

The deans helped in raising financial support for the proposed foundation and in arguing the case before the regents. Their major point was that a program of patent management was essential to protect the public from exploitation by pharmaceutical companies. The regents remained uncertain about the wisdom of implementing the plan at Wisconsin, but gave their reluctant endorsement when it was clarified that the university would have no responsibility, financial or otherwise, for the foundation (University of Wisconsin 1925).

The regents' approval precipitated opposition to the plan from within and outside the university. Faculty members of the College of Agriculture expressed concern that the plan might initiate the dangerous trend of diverting university research from the goal of aiding farmers to that of making money through patentable ideas.[31] The editor of *Hoard's Dairyman* threatened to publish stinging indictments to the effect that scientists have no right to patent discoveries made on public time and financed by public funds, that such discoveries should be made freely available to the public, and that awarding the patents to a private foundation or corporation only compounds the injury to the public. Some scientists nationwide reiterated their view that it was wrong in principle to patent scientific discoveries.

In the face of such criticism, the foundation was incorporated on 5 November 1925 as the Wisconsin Alumni Research Foundation (WARF), with a trusteeship comprised of wealthy Wisconsin graduates.[32] Supporters of WARF went on the counteroffensive. Russell assumed the role of spokesperson, attacking the critics for their naiveté in pushing a policy of donation of patent rights to the public. He argued that such a course only opened the door for misuse and fraud. The critics were also accused of perpetuating the image of corporate activity as synonymous with exploitation.[33]

WARF's first contract in July 1927 provided Quaker Oats with an exclusive license to use the irradiation process in cereal foods (*New York Sun* 1927). When the patent was officially granted in August 1928, the way was clear to widen the foundation's commercial base. Several firms were licensed in 1929 to market concentrated vitamin D oil. The earnings were substantial, amounting to about a thousand dollars a day in royalties from these and other patent licenses. It became clear that WARF needed a full-time manager to handle all the transactions, and in 1930 Russell resigned his deanship at the university to take the job (Beardsley 1969).

By 1931 WARF had amassed an endowment of $400,000. When news of this reached the press, the foundation was accused of collaborating with big business to exploit the public and of using royalty income not to help the university but to purchase corporate stocks and build enormous capital reserves (*Madison Capital Times* 1931). Russell responded that the endowment had to be built up before the foundation could consider dispersing royalties to university activities. Within the year, however, WARF established a program of doctoral fellowships at the university. In 1933, in the depths of the Depression, a grant of more than $250,000 was awarded to the university, and a chair of wildlife management was donated to the College of Agriculture (Russell 1935). WARF increasingly favored support of studies on practical applications related to the patents they held. Through WARF grants and fellowships, the university developed the largest graduate biochemistry program in the United States. By 1950 the Wisconsin program accounted for 30 percent of the biochemistry Ph.D. graduates in the country annually.

WARF's public image continued to decline, however. Public health professionals argued that WARF's monopolistic control of the patent, and its licensing fees, had raised the price of vitamin D irradiated milk (*Journal of Pediatrics* 1936). Other health-related inventions handled by WARF—the Hart patent on copper-iron salts for treatment of anemia (1932) and the Hisaw-Fevold patent on pituitary hormones (1936)—were also a source of concern. Questions were raised, too, about possibly mercenary motivations on the part of WARF and even Steenbock himself.[34] WARF's image was further damaged when the foundation began in 1937 to bring a series of suits for infringement of the Steenbock and Hart patents. The cases were complicated, resulting in appeals and counterappeals over the next ten years.[35]

With these costly, sometimes ugly, court battles as background, the press called for a probe of WARF's finances, and the federal government threatened antitrust proceedings against WARF. In testimony before a congressional subcommittee in 1943, the foundation was accused of abusing its patent rights. There were charges that WARF (1) was not interested in research unless a commercial advantage could be obtained from it; (2) suppressed the use of competing processes; (3) suppressed research data that were at variance with its monopoly interest; (4) discouraged research by firms holding licenses under its licensing scheme; (5) attempted to suppress or prevent truthful advertising to eliminate competition; and (6) collaborated to maintain artificially high prices for vitamin D irradiated products. WARF's defense of its policies did not have much impact on the public's perception of the situation (U.S. Congress, Senate 1943). The foundation argued that far from operating as a "drug combine," it was acting as a not-for-profit foundation to protect the public; that state and federal codes rather than royalty charges had forced up the cost of products irradiated to enhance vitamin D content; and that the patents stimulated widespread use of the process and secured funds for other valuable research (Ross and Schoenfeld 1948).

The government, the courts, and the public were not convinced by these arguments. Several legal decisions went against WARF, and press and government criticism of WARF's ethics continued into the 1960s and 1970s (Maisel 1948; *Madison Capital Times* 1931).[36] WARF, meanwhile, stood firm on the elements of its earlier self-justification. The University of Wisconsin, caught between the financial advantages of its relation to WARF and the scruples it was expected to exercise as a publicly funded and publicly oriented institution, underwent frequent rounds of discussion on both sides of the issue.

The WARF experience sent a message to many other universities that more was at stake in patenting than financial returns. In most instances, patents did not pay off for universities; when they did, problems of public credibility and possible conflicts of interest came to the fore. After World War II, massive federal funding of campus research spawned potentially lucrative applications, and many universities began to face similar dilemmas. The postwar problems were highlighted in a provocative report prepared for the U.S. Senate in the late 1950s.

The Melman Study

Early in 1956, Seymour Melman, professor of industrial engineering at Columbia University, was asked by the Senate Subcommittee on Patents, Trademarks, and Copyrights for help in its far-reaching review of the patent system. Melman was asked to focus on the industrial experience—a sphere in which he had considerable expertise. He agreed to undertake a study for the committee (Melman 1958).

Melman began by discussing the project with people in his own institution. These in-house discussions at Columbia University led him to select four major companies—Bell Telephone Laboratories, Standard Oil of New Jersey, Du Pont, and Ford Motor Company—as the focus of his study. The companies selected appeared to be representative, covering a wide range of industries—communications, oil, chemicals, and transportation—and involving a major portion of research and development effort where patents came into play. Melman broadened the scope of his study to include university laboratories, especially as it became evident that there were concerns about patents in the academic environment that needed to be addressed despite the apparent consensus that the problems were fewer than in industry. Among the academic institutions he chose to include in the survey were the Massachussets Institute of Technology, Columbia, New York University, Rockefeller Institute, Purdue, and Rutgers.[37]

Melman visited companies and university laboratories and conducted interviews. It was not always clear who would be the best people to interview. The obvious candidates were in an institutions's patent department, but they often turned out to be so absorbed in the technical details of the patent application process that they seemed unable (or unwilling) to open their minds to the larger questions. Melman's most fruitful discussions were with research directors, who tended to be more attuned to the policy effects of the interrelation of patents and research.

The evidence collected in the survey suggested not only that there were problems with the way the patent system was running, but also that there were serious questions about the viability of the system per se. The system was based on a *single-inventor* concept, whereas modern science and technology had become a group-oriented endeavor in which it was often difficult to identify the individual or individuals responsible for a discovery or invention. No survey was really required to highlight this problem; a notorious case at the time—the streptomycin case at Rutgers—was fraught with litigation seeking to expand inventorship recognition. The case illustrated a basic source of incompatibility between the patent system and the nature of the research process.

Melman's survey identified other such conflicts, some of which were already recognized to some degree, and went even further by undercutting the arguments that proponents of the system had been using to outbalance the perceived disadvantages. One commonly held view was that patents acted as competition and thus provided a unique impetus to research. Melman's study found that patents were only one tactic used in competition, and that others would remain or emerge even if there were no patent system. Furthermore, the competition generated by patents centered on the building up of enormous financial reserves that were subsequently squandered in legal confrontations rather than used in a positive way to stimulate further research and discovery.

In the course of his study, Melman grew to be more interested in the question of whether the system was living up to its constitutional mandate to encourage discovery and invention than in the viability of the patent system (that is, how it worked). This question was not something he had thought about before embarking on the study, but it emerged while he worked through the data and reflected on the various issues. Moreover, it represented a more fundamental assault on the system than he had originally conceived or than the subcommittee expected him to undertake. He was aided in this effort by what he perceived as his relatively open view about the concept of intellectual property rights; he did not come to the project with an assumption that such rights were fundamental and inviolable, and he was willing to entertain the notion that they were limited.

In Melman's view the patent system perpetuated a "fearful array of struggles" over property rights that reduced rather than encouraged further research. This conclusion was clear from his data on industry, but it was also painfully evident from his study of academic environments. Pressures toward a policy of aggressive patenting had arisen in educational institutions eager to tap new sources of research funding. The policy was backfiring, however, because increasing managerial control of research undertakings impeded the university scientist in the free pursuit of knowledge as an end in itself and thus weakened the universities as centers of basic research. Another problem was secrecy. Prompt publication, an essential aspect of the process of scientific inquiry (especially in the integrated, collective nature of contemporary research practice), was being eroded because of the secrecy requirements of the patent application process.

Academic scientists were willing to express their views on this quite freely, in part, Melman speculated, because it was a relative novelty to find the university becoming more structured around patents and straying more from the goal of seeking knowledge for its own sake. The issue was extremely sensitive; it was causing major rifts among faculty and staff. One professor at a major institution had become so frustrated by the situation that he accused his colleagues of worshiping the golden calf, and was officially censured for speaking out.

Melman analyzed several cases in which patents were believed to be distorting the research aims and priorities in the academic environment. As an experiment, he cited the cases anonymously in interviews and asked people to try to identify the institution involved. The wide array of responses suggested to Melman that the problems connected to these cases were hardly unique; in fact, they were quite common to the academic community as a whole. That the problems were sensitive and controversial was evident from the nature of the responses, coupled with the "golden calf" incident. In preparing his final report, Melman found himself having to remain circumspect about his sources. He was "spilling a lot of beans" merely by describing the situation, and to have named names would have opened up the possibility of ugly recriminations.

Melman's report concluded that the patent system was "a restraint rather than an impetus" to research; he recommended that the system be terminated and replaced by one based on nonmonetary awards. The report was submitted to the subcommittee toward the end of 1957. A "long silence" ensued. It was clear that the subcommittee members and staff were somewhat taken aback by the analysis and conclusions. They had expected Melman to offer suggestions about improving the system, not dismantling it. When they did finally respond, they asked him to modify certain statements. He agreed to some changes but refused to touch the conclusion. Eventually he told them to publish the report as it stood or not at all—just as they saw fit. Tactfully, they agreed to publish with an introduction making the appropriate disclaimer that the views presented in the report did not reflect the views of the subcommittee.

The report was published in 1958. Predictably, there was a negative response from the patent lawyers, who had a great deal to lose if Melman's recommendations were taken seriously. One behalf of the Patent Law Association of Los Angeles, Richard F. Carr prepared a rebuttal (Carr 1960). The paper used the results of a survey of users of the patent system to undercut Melman's findings and to defend the integrity and viability of the patent system as it operated.

According to Carr, his survey suggested that the patent system had consistently served as a stimulus to science and invention. This conclusion was diametrically opposite from that of the Melman report. One reason for this difference, Carr alleged, was in the statistical shortcomings of Melman's study. Melman had restricted his survey to large industrial organizations and to individual scientists from those organizations and from a few university and government laboratories. He seemed to be laboring under the false assumption that patenting was of primary significance in big business. However, Carr's survey—based, he claimed, on a larger, more representative statistical sampling—indicated the importance of patenting in a range of organizations. Patenting, it turned out, was even more important to the economic well-being of smaller organizations, which had less capital to rely on. Carr felt that Melman had neglected such considerations and had focused disproportionately on, for instance, university scientists, whose work by its very nature rendered them far less attuned to the importance of the patent system.

The paper attempted to refute, often with scathing remarks about Melman's historical accuracy and intellectual grasp of the patent system, each of the conclusions highlighted in the Melman report. Melman had stated that the patent system was set up to serve individual inventors, whereas modern science and technology had developed into a group effort in which the concept of individual inventorship was no longer applicable. In response, Carr stressed the numerous provisions in the system for group patenting. He also suggested that Melman may have been confused by the nature and size of the group; although many scientists and technicians worked on a project together, it was generally the case that only one or two *conceived* the project in a patentable sense. Furthermore, Carr accused Melman of perpetuating a historical myth, of not being sufficiently sensitive to the continuum that existed between the situation of nineteenth- and twentieth-century scientists, and their relation to the patent system.

Carr took as much exception to the Melman report's warnings about the dangers of patenting as to its remarks about the obsolescence of the system. Here the focus was on secrecy. Contrary to Melman's survey, Carr suggested that—far from increasing the likelihood of secrecy—patenting in fact eliminated the need for it. A patent represented a full disclosure of how an invention works. Furthermore, with the protection from marketable use provided by a patent, inventors did not need to worry so much about releasing information, and the result was a freeing up of the process of information flow between scientists and laboratories. Carr felt that Melman's speculations about the greater openness of a patent-free system were ill-founded and that instead there would inevitably be greater caution and a higher degree of secrecy.

The patent attorney argued that basic research was not an isolated activity, and that the only real justification for it was as a means of evolving usable scientific knowledge. The patent system had served the public well in bringing basic research to the point of usability. Without the system, there would still be pressure to produce usable knowledge—but without the control, income for further research, and other advantages that the system provided.

Predictably, Melman's report offended the patent professionals; however, the critique they offered did not refute his data. The Senate subcommittee, while not endorsing Melman's recommendations, described his report as "thoughtful and competent," and as a "valuable contribution to the literature." Although his study did not lead to major changes in the patent system or in university policies, the problems he raised have reappeared in different contexts in the past quarter century.

The Controversy Continues

The biotechnology boom of the 1980s has brought these unresolved problems to the fore once again. The new practical potential of academic biology has been realized in an economic and political environment that encourages universities to become more closely involved with industry through patenting, research contracts, and joint endeavors. More and more universities are adopting aggressive patenting policies. As they do, they are experiencing many of the same problems seen in the past—unfulfilled expectations, conflicts of interest, wavering public credibility, communication barriers in research, and—especially in biomedical applications—ethical dilemmas. As in the past, there is little serious and sustained attention being given to alternatives that can enhance the university's education, research, and public-service functions while avoiding the pitfalls involved in patenting.

Notes

Acknowledgment: The historical cases in this article were researched and prepared in collaboration with Philip Alexander. The research was supported by the National Science Foundation (History and Philosophy of Science Program), under Grant SES-8510448, and the National Endowment for the Humanities (Humanities, Science and Technology Program), under NEH Grant RH-2076686

1. See, for example, Hale (1921), Sevringhaus (1932), Gregg (1933), Henderson (1933), Gray (1936), Connolly (1937), and American Medical Association (1939). See also American Chemical Society (1937), and American Association for the Advancement of Science (1934).

2. For my previous discussion of the California and Wisconsin cases, see Weiner (1982, 1986). The latter also includes an account of the Melman study and of contemporary patent problems in biotechnology.

3. Biographical sketches of Robertson include Marston (1932), Wasteneys (1930), Ostwald (1930), Tiegs (1930, 1951), and McGuire (1970).

4. The relationship between Robertson and Cottrell, as well as the development of Robertson's early interest in patenting, is documented in the Diaries of F. G. Cottrell, Manuscripts Division, Library of Congress. A detailed search of archival records at the University of California–Berkeley has yielded no evidence that Cottrell's patent was offered to the university or that it was discussed by the regents.

5. Robertson to Loeb, 16 February 1911, 12 March 1911, 25 October 1915, 28 December 1916; Loeb to Robertson 3 January 1912, 20 October 1915, 31 December 1915,

13 May 1916; in Jacques Loeb Papers, Box 13, Manuscripts Division, Library of Congress.

6. A. O. Leuschner to B. I. Wheeler, 21 April 1917; "Outline of a proposed agreement between the Regents of the University of California and T. B. Robertson," April 1917; W. Olney, Jr., to Wheeler, 8 May 1917; Robertson to Wheeler, 24 May 1917; in President's Papers, Archives, University of California–Berkeley.

7. Robertson to F. G. Cottrell, 26 June 1917; Cottrell to Robertson, 19 July 1917; in Records of the Office of the Secretary (temp. ser.), 1912–1959, Smithsonian Institution Archives.

8. Small amounts of income from tethelin totaling less than $300 were noted in the *University of California Bulletin* (1920, 1922, 1924).

9. Loeb to H. Wasteneys, 19 April 1917, 22 October 1917; in Jacques Loeb Papers, Box 15, Manuscripts Division, Library of Congress.

10. R. Falconer to Robertson, 15 March 1918, 30 March 1918; Robertson to Falconer, 1 April 1918; J.J.R. Macleod to Falconer, 23 February 1918; in Office of the President (Falconer), A67-0007/049–050, University of Toronto Archives.

11. Robertson to Loeb, 8 February 1920, 9 February 1921, 22 January 1924; in Loeb Papers, Box 13, Manuscripts Division, Library of Congress. Robertson's Australian career is well documented in collections at the Archives Division, State Library of South Austrialia, Adelaide; Special Collections, University of Adelaide, and the CSIRO Archives, Canberra.

12. For biographical information, see *National Cyclopedia of American Biography* (n.d.) and Rubin (1980).

13. G. W. McCoy to G. F. Dick, 12 November 1924, quoted in American Medical Association (1937).

14. For a summary of the thyroxin affair, see American Medical Association (1918).

15. See also American Medical Association (1927).

16. Dick et al. v. Lederle Antiotoxin Laboratories, 43 F.2d 628-40 (S.D.N.Y. 1930).

17. G. H. Dick to Editor, Science Press, 15 March 1933; Parkinson and Lane to Science Press, 25 April 1933; in A. Gregg Papers, Box 23, National Library of Medicine.

18. [?] to Gregg, 10 May 1932 [1933?]; S. Flexner to Gregg, 2 May 1933; in Alan Gregg Papers, Box 23, National Library of Medicine. See also Sexton (1967).

19. W. M. Fletcher to Gregg, 10 April 1933, in Alan Gregg Papers, Box 23, National Library of Medicine.

20. M. C. Williams to Gregg, 17 December 1951, in Alan Gregg Papers, Box 24, National Library of Medicine.

21. For a good review article on developments in the research, see Minot and Isaacs (n.d.). For biographical information on Cohn, see *National Academy of Sciences* (1961); for Minot, see Castle (1952) and Rackemann (1956).

22. For A. L. Lowell to D. L. Edsall, 14 March 1927, see quotation in Cohn (1951, p. 5).

23. E. J. Cohn to O. Roberts, 12 April 1927 (Cohn, p. 5).

24. Cohn to E. M. Berolzheimer, 31 January 1929 (Cohn, pp. 6–7).

25. F. G. Cottrell to A. H. White, 12 April 1927 (Cohn, pp. 6– 7).

26. H. Dale to [?], [1929], in Records of the Committee on Pernicious Anemia, Rare Books Division, F. A. Countway Library, Harvard University.

27. Correspondence in "George R. Minot and others re anemia 1927," folder in E. J. Cohn Papers, F. A. Countway Library, Harvard University.

28. W. B. Cannon to President J. B. Conant, 18 January 1935, in W. B. Cannon Papers, F. A. Countway Library, Harvard University.

29. For an overview of Steenbock's life and work, see Schneider (1973). See also Steenbock (n.d.)

30. Steenbock to J.J.R. Macleod, 14 February 1925; Steenbock to E. C. Kendall, 14 February 1925; Macleod to Steenbock, 18 February 1925, in Steenbock Collection, General Files, Correspondence 1925–1929, University of Wisconsin Archives.

31. F. B. Morrison to H. L. Russell, 28 October 1925, in WARF Files, University of Wisconsin Archives.

32. For unpublished reviews of WARF's history, see Culotta (1968) and Cohen (1971). See also Fred (1973). More recent accounts and interpretations include Weiner (1982, 1986), and Blumenthal, Epstein, and Maxwell (1986).

33. H. L. Russell to F. B. Morrison, 22 January 1926, in WARF Files, University of Wisconsin Archives.

34. A. Gregg to Sir W. Fletcher, 3 March 1932, in Alan Gregg Papers, Box 23, National Library of Medicine.

35. See WARF v. Vitamin Technologists, 1 FRD 8 (S.D. Cal. 1939); Vitamin Technologists v. WARF and WARF v. Vitamin Technologists, 146 F.2d 941 (9th Cir. 1945). See also American Medical Association (1945).

36. Criticism also found in [?] to F. H. Harrington, 2 June 1965, in Presidents' Papers (Harrington), General Correspondence, 1964–1965, Box 87, University of Wisconsin Archives.

37. The following information and quotes are from an interview by the author with Seymour Melman, 19 April 1985, and notes in Melman's files.

References

American Association for the Advancement of Science. 1934. "The Protection by Patents of Scientific Discoveries." Report of the Committee on Patents, Copyrights, and Trade Marks. *Science* 79 (January supplement): 7–40.

American Chemical Society. 1937. "Are Patents on Medical Discoveries and on Foods in the Public Interest?" Joint Symposium Presented before the Divisions of Medicinal Chemistry, Biological Chemistry, and Agricultural Food Chemistry. *Industrial and Engineering Chemistry* 29 (November): 1315–1326.

American Journal of Public Health. 1926. "Ethics and Patents." *American Journal of Public Health* 16: 919–920.

American Medical Association. 1918. *Report of the Judicial Council*. Sixty-ninth Annual Session, Chicago, 10–14 June 1918. American Medical Association Archives.

———. 1927. "Current Comment: The Scarlet Fever Patents." *Journal of the American Medical Association* 88 (23 April): 1324.

———. 1937. "Correspondence: The Scarlet Fever Patents." *Journal of the American Medical Association* 109 (27 November): 1833.

———. 1939. Conference on Medical Patents. *Journal of the American Medical Association* 113 (22, 29 July): 327–336, 419–427.

———. 1945. "Wisconsin Alumni Group Fights to Validate Vitamin D Patents." *Journal of the American Medical Association* 128 (5 May): 38.

Beardsley, E. H. 1969. *Harry L. Russell and Agricultural Science in Wisconsin.* Madison: University of Wisconsin Press, esp. chap. 11 ("Making Research Pay Its Own Way"), pp. 155–171.

Blumenthal, D., S. Epstein, and J. Maxwell. 1986. "Commercializing University Research: Lessons from the Experience of the Wisconsin Alumni Research Foundation." *New England Journal of Medicine* 314 (19 June): 1621–1626.

British Medical Journal. 1927a. "Patented Research." *British Medical Journal* 1 (12 March): 479–480.

———. 1927b. "Patented Research." *British Medical Journal* 1 (19 March): 526.

———. 1927c. "Patented Research." *British Medical Journal* 1 (14 May): 881–882.

Carr, R. F. 1960. "Our Patent System Works." *Journal of the Patent Office Society* 47 (May): 295–326.

Castle, W. B. 1952. "The Contributions of George Richard Minot to Experimental Medicine." *New England Journal of Medicine* 247: 585–592.

Cohen, E. B. 1971. "The House That Vitamin D Built: The Wisconsin Alumni Research Foundation." Report prepared for the Center for a Responsive University. WARF Files, University of Wisconsin Archives.

Cohn, E. J. 1951. *History of the Development of a Patent Policy Based on Experiences in Connection with Liver Extracts and Blood Derivatives, 1927–1951.* University Laboratory of Physical Chemistry Related to Medicine and Public Health, Harvard University.

Congressional Record (Senate). 1965. "Private Patent Monopolies." (17 May): 10, 343–344.

Connolly, A. 1937. "Should Medical Inventions Be Patented?" *Science* 86 (29 October): 383–387.

Culotta, C. A. 1968. "Research and Funding at the University of Wisconsin, 1914–1933." WARF Files, University of Wisconsin Archives.

Daily Californian. 1917. "Dr. Robertson Isolates and Discovers Tethelin." *Daily Californian,* 15 January.

Dick, G. F., and G. H. Dick. 1924a. "The Etiology of Scarlet Fever." *Journal of the American Medical Association* 82 (26 January): 301–302.

———. 1924b. "A Skin Test of Susceptibility to Scarlet Fever." *Journal of the American Medical Association* 82 (26 January): 265–266.

———. 1925. "Scarlet Fever Preparations." *Journal of the American Medical Association* 85 (26 September): 996.

———. 1927. "The Patents in Scarlet Fever Toxin and Antitoxin." *Journal of the American Medical Association* 88 (23 April): 1341–1342.

Dick, G. H. 1939. "A Brief History of the Scarlet Fever Patent." Conference on Medical Patents. *Journal of the American Medical Association* 113 (22 July): 327–330.

Fred, E. B. 1973. "The Role of the Wisconsin Alumni Research Foundation in Support of Research at the University of Wisconsin." Wisconsin Alumni Research Foundation.
Gray, G. W. 1936. "Science and Profits." *Harper's* 172: 539–549.
Gregg, A. 1933. "University Patents." *Science* 77 (10 March): 257–259.
Hale, W. J. 1921. "University Researchers Should Patent Discoveries in Their Own Names." *Chemical and Metallurgical Engineering* 25 (16 November): 913–914.
Henderson, Y. 1933. "Patents Are Ethical." *Science* 77 (31 March): 324–325.
Journal of Pediatrics. 1936. Report of the Committee on Clinical Investigation and Scientific Research. *Journal of Pediatrics* 8: 124–130.
Lancet. 1927a. "The Patenting Question." *Lancet* 1 (12 March): 551.
______. 1927b. "The Patenting Question." *Lancet* 1 (26 March): 674–675.
McGuire, M. 1970. "Personalities Remembered." A talk for the Australian Broadcasting Commission, 20 September.
Madison (Wisconsin) *Capital Times*. 1931. *Madison Capital Times*, 21 May, 2 July.
______. 1965. "Senator Long Shows That Private Firms Get Patents from Public Research." *Madison Capital Times*, 21 June.
______. 1971. "WARF Investments Ignore Social Concerns." *Madison Capital Times*, 12 May.
Maisel, A. Q. 1948. "Combination in Restraint of Health." *Reader's Digest* (February): 42–45.
Marston, H. R. 1932. "Thorburn Brailsford Robertson." *Australian Journal of Experimental Biology and Medical Science* 9: 1–5.
Melman, S. 1958. *The Impact of the Patent System on Research: Study No. 11 of the Subcommittee on Patents, Trademarks, and Copyrights of the Senate Committee on the Judiciary*. 85th Cong., 2d sess. Washington: Government Printing Office.
Minot, G. R., and R. Isaacs. N.d. "Pernicious Anemia: Review of the Progress in the Study of the Disease in North America during 1928." MS, R. Isaacs Papers, Box 2, Rare Books Division, F. A. Countway Library, Harvard University.
National Academy of Sciences. 1961. "Biographical Memoirs." *National Academy of Sciences* 35: 47–84.
National Cyclopedia of American Biography. N.d.
"Dick, George Frederick." *National Cyclopedia of American Biography* 54: 240.
______. N.d. "Dick, Gladys Rowena Henry." *National Cyclopedia of American Biography* 51: 107.
New York Sun. 1927. *New York Sun*, 17 July.
New York Times. 1917. "Man May Be Self-Made." *New York Times*, 6 January.
Ostwald, J. B. 1930. "Prof. T. Brailsford Robertson." *Nature* 125 (15 February): 245.
Rackemann, F. M. 1956. *The Inquisitive Physician: The Life and Times of George Richard Minot*. Cambridge: Harvard University Press.
Robertson, T. B. 1916a. "On the Isolation and Properties of Tethelin, the Growth-Controlling Principle of the Anterior Lobe of the Pituitary Body." *Journal of Biological Chemistry* 24: 409–421.
______. 1916b. "The Effects of Tethelin: Acceleration in the Recovery of Weight Lost during Inanition and in the Healing of Wounds." *Journal of the American Medical Association* 66 (14 November): 1009–1011.

———. 1917. "The Utilization of Patents for the Promotion of Research." *Science* 46 (19 October): 371–379.

Ross, W., and C. Schoenfeld. 1948. WARF Report. *Wisconsin Alumnus* (June): 21–31.

Rubin, L. P. 1980. "Dick, Gladys Rowena Henry." In *Notable American Women: The Modern Period,* edited by B. Sicherman, Carol Hurd Green, Ilene Kantrov, and Harriette Walker, pp. 191–192. Cambridge: Belknap Press of Harvard University Press.

Russell, H. L. 1935. *A Decade of Service, 1925–1935: Report of the Director, Wisconsin Alumni Research Foundation*. WARF Library.

Schneider, H. A. 1973. "Harry Steenbock (1886–1967): A Biographical Sketch." *Journal of Nutrition* 103 (September): 1235–1247.

Sevringhaus, E. J. 1932. "Should Scientific Discoveries Be Patented?" *Science* 76 (9 September): 233–234.

Sexton, A. M. 1967. *A Chronicle of the Division of Laboratories and Research, New York State Department of Health: The First Fifty Years, 1914–1964*. Lunenburg, Vt.: Stinehour, pp. 104–109.

Steenbock, H. 1924a. Manuscript concerning general background of Steenbock patent as submitted to A. B. Marvin, Patent Attorney. WARF Files, University of Wisconsin Archives.

———. 1924b. "The Induction of Growth Promoting and Calcifying Properties in a Ration by Exposure to Light." *Science,* n.s. 60 (5 September): 225.

———. 1927. "The Invention of Antirachitic Process and Products." WARF Files, University of Wisconsin Archives.

———. N.d. "The Relation of the Writer to the Wisconsin Alumni Research Foundation and the Events Which Led to Its Organization." WARF Files, University of Wisconsin Archives.

Tiegs, O. W. 1930. "Thorburn Brailsford Robertson." *Nature* 125 (15 February): 245.

———. 1951. "Thorburn Brailsford Robertson." *M.S.S. Review: A Journal of the Adelaide Medical Students* (May): 21–23.

U.S. Congress. Senate. 1943. Subcommittee on Scientific and Technical Mobilization of the Committee on Military Affairs. *Vitamin D: Monopoly and Cartel Practices*. 78th Cong., 1st sess., hearings pursuant to S. Res. 107.

University of California Bulletin. 1920. "Annual Report of the President of the University." *University of California Bulletin,* 3d ser., 14 (December): 305.

———. 1922. "Annual Report of the President of the University." *University of California Bulletin,* 3d ser., 16 (December): 337.

———. 1924. "Annual Report of the President of the University." *University of California Bulletin,* 3d ser., 17 (January): 280.

University of Wisconsin. 1925. Excerpt from minutes of Executive Committee, Board of Regents, University of Wisconsin, Madison, 8 May. Steenbock Collection, WARF Files, Misc. to 1925, University of Wisconsin Archives.

Wasteneys, H. 1930. "Thorburn Brailsford Robertson (1884–1930)." *Biochemical Journal* 24: 577–578.

Weiner, C. 1982. "Science in the Marketplace: Historical Precedents and Problems." In

From Genetic Experimentation to Biotechnology: The Critical Transition, edited by W. J. Whelan and S. Black, pp. 123–131. New York: John Wiley and Sons.

———. 1986. "Universities, Professors, and Patents: A Continuing Controversy." *Technology Review* 89 (2): 33–43.

6. Biotechnology, Plant Breeding, and Intellectual Property: Social and Ethical Dimensions

Frederick H. Buttel
Jill Belsky

The past half-dozen years have witnessed in many countries the mobilization of public interest groups that seek to reverse global trends toward proprietary protection of plants and plant parts. Numerous articles have appeared in the popular, trade, and science press concerning the benefits and costs of plant variety protection and plant patenting. Private-sector seed and agricultural biotechnology companies have found it necessary to defend the social desirability and ethical neutrality of proprietary protection to a degree that would be uncommon for other types of industries with product lines protected by comparable intellectual property arrangements.

Although the scope and intensity of current debates over proprietary protection of plants are unprecedented, there has actually been a long history of struggle in the United States and other industrial countries over the means by which private firms could profitably market and sell plants (Kloppenburg 1985). Indeed, there are a number of similarities between the struggles during the first three decades of the twentieth century and those which have occurred since the late 1970s. As we will argue below, the conditions for a profitable private seed industry have continued to revolve around reducing or eliminating competition from three sources: (1) farmers, who may save their own seed for planting the next cropping season; (2) the public research system, which has historically developed new and improved crop varieties; and (3) other seed companies.

In this chapter we provide a brief overview of the development of the seed industry in the United States, particularly in relation to public plant breeding institutions that have both supported and competed with private-sector efforts. We then discuss major types of intellectual property arrangements that pertain to private plant breeding and identify several crucial issues in proprietary protection of plant breeding inventions.

Agriculture, Seeds, and Plant Breeding

The seed industry and the larger agricultural sector have some characteristics that set them apart from other industrial sectors. Most important for our purposes, these differences between the seed industry and most manufacturing industries have dictated particular strategies by seed companies to establish the conditions for profitable private plant breeding. It is useful to discuss these particularities of the agricultural seed industry in order to understand better the context of intellectual property arrangements.

The basis of plant agriculture is plant propagation. Plants must be propagated by seeds, cuttings, bulbs, or some other method in order to grow a crop. In nature, most economically relevant characteristics are inherited, and plant progeny tend to be genetically similar to their parents. To improve the genetic potential of their crops over time, farmers themselves can select for superior phenotypes and thereby improve crop performance. Thus, there are certain natural barriers to the establishment of seeds and other plant propagation products as commodities to be produced by companies on a privately profitable basis. (Hereafter we refer only to seeds rather than to such other planting material as tubers or cuttings.) First, seed companies must be able to develop varieties that are better than those which farmers can breed through recurrent selection or introduction of exotic germplasm. Second, and most important, the genetic similarity of successive generations of crop varieties enables farmers to save their own seeds for the next cropping season; even if a private company develops a qualitatively superior variety, the ability of farmers to save their seeds would imply little long-term market potential for the variety.

There is a technical solution to the internal obstacle to commercial enterprise posed by these integral features of the seed: hybridization. The significance of hybridization is twofold. First, through the biological phenomenon of heterosis (or "hybrid vigor"), a successful hybrid will outyield conventional open-pollinated varieties. Second, the progeny of hybrid seeds are genetically dissimilar from the parents, and their yield performance reverts to a level significantly lower than that of the parents by some 15 to 40 percent. Thus, farmers who use hybrid seeds must enter the market each growing season in order to purchase new seeds, thereby creating the market potential for a profitable seed business.[1] If hybrid varieties are clearly superior to nonhybrids, economic forces will in a short time make the use of hybrid seeds obligatory for all farmers, further expanding the market.

Another peculiarity of the agricultural sector and the seed industry is the strikingly large presence of publicly funded agricultural research and development. The farm sector in advanced market economies tends to have a decentralized ownership structure so that virtually no individual producers have the resources to perform their own research. Thus, the bulk of agricultural research in the United States and other Western countries has historically been performed by the state.[2]

Probably the single most important contribution of public agricultural research to increased farm productivity has been that of plant breeding and such auxiliary activities as germplasm collection and evaluation. From the middle of the nineteenth century through the early years of the twentieth century, the U.S. Department of Agriculture engaged in major programs of global germplasm collection and freely distributed seed of improved varieties to farmers. From the early twentieth century on, plant breeders in the land-grant universities (LGUs) and what is now known as the Agricultural Research Service (ARS) have developed "finished" crop cultivars, most of which have been released as public-domain varieties to seed improvement associations and to all interested seed companies. The role of public plant breeding institutions in developing new, improved varieties provides significant competition to private seed companies. The historic solution to this "social obstacle" to capital accumulation by private industry has been political in nature; pressure has been placed on the public agricultural R and D system to refrain from releasing finished varieties in crops considered to be privately profitable by seed companies and generally to move public research away from "applied" activities to more "basic" work complementary to, rather than competing with, private research.[3]

There is also a juridical solution to eliminating the three forms of competition: the extension of proprietary rights to plant germplasm. Intellectual property statutes enable an individual seed company to develop new knowledge and products that can be denied to competitors. Thus, a seed company will have a greater incentive to develop new plant varieties than would be the case if there were no intellectual property restrictions. There are important interactions between biological (hybridization), political (elimination of public competitors), and legal routes for achieving profitable plant breeding in the private sector. Hybridization, for example, not only eliminates competition from farmers, but also reduces the need for legal protection since hybrids have a biological patent because parental lines can be maintained as trade secrets. Also, to the extent that expanded property rights encourage apparent increases in private research, this route provides a lever to shift the social division of labor to a balance more favorable to private interests. Nonetheless, intellectual property restrictions have recently become increasingly important in the seed industry, especially as it has experienced major structural changes since the 1970s.

Efforts to achieve intellectual property goals in plant germplasm have a long history. Seed companies and the American Seed Trade Association (ASTA) had lobbied intensely for effective intellectual property restrictions on plant breeding inventions for some years before the passage of the Plant Patent Act of 1930. Further political efforts on the part of the private sector led to the Plant Variety Protection Act of 1970 and its 1980 extensions. Following the Supreme Court's landmark decision in Diamond v. Chakrabarty (447 U.S. 303 [1980]), the Board of Patent Appeals in its Ex parte Hibberd ruling of 1985 has determined that it is possible in the United States to protect plant varieties under the General Patent Law (section 101).

Changes in legal arrangements have been stimulated further by the potential of powerful new genetic technologies, which together are referred to as *biotechnology*. Biotechnology has been hailed as an epochal new technology that is expected to undergird a new regime of commercial expansion. The private agricultural input industry is gearing up to exploit a market for genetically engineered plant varieties projected to be worth $7 billion by the year 2000. Nevertheless, the commercial promise of these new legal and technical conditions makes it critical that the processes of private appropriation of plant breeding inventions be examined so that socioeconomic and ethical concerns can be addressed.

A Brief History of U.S. Plant Breeding[4]

Most of the major crops grown in the United States are not native to North America. An important factor in U.S. agricultural settlement was the importation of new crops that would perform adequately in the ecological and socioeconomic environments of the emerging nation. Major efforts along this line were undertaken by the U.S. government and by private agricultural societies during the eighteenth and early nineteenth centuries. With the establishment of the USDA as a cabinet-level agency and the founding of the LGU system in the early 1860s, germplasm collection and dissemination became the responsibility of USDA and LGU agronomists, botanists, and plant breeders. "As late as 1878 fully a third of the Department's annual budget was spent on germplasm collection and distribution" (Kloppenburg 1985).

Two major factors led to the slow diminution of the USDA role in germplasm distribution during the late nineteenth and early twentieth centuries. First, with the passage of the Hatch Act in 1887, which authorized federal funding for state agricultural experiment stations (SAESs), LGU plant research was greatly strengthened.[5] Second, during the middle of the ninteenth century there emerged a nascent private seed trade that revolved primarily around the sale of vegetable and flower seed to commercial and home gardeners.[6] These private companies, however, hoped to be able to penetrate the market for field crop seed, but found the public role in seed distribution a formidable market barrier. ASTA was formed in 1883 with the motivation of influencing government policy. Its first major victory came in 1924 when it was able to persuade Congress to end free federal distribution of seeds.

But the termination of this federal largesse left in place an extensive public plant breeding infrastructure in the LGU/SAES system. By the time that free seed distribution had been terminated, plant breeding was moving rapidly toward a scientific rather than craft basis. During the first quarter of the century, LGU plant breeders released a growing number of new varieties, thereby providing new competition for private seed companies. The seed companies responded through ASTA by establishing a Committee on Experiment Stations to monitor their activities and to foster a more favorable division of labor with the private sector.

The most dramatic achievement of public breeding in the 1920s and 1930s was the successful hybridization of corn. Koppenburg (1985) has noted that

> ironically, it was this very success that ensured that plant breeding would not long remain the more or less exclusive province of the public sector. Nor would public breeders be able to maintain a "strictly disinterested attitude" toward their work. For [various] reasons hybridization made corn breeding privately profitable. In 1926, Henry A. Wallace founded the first company devoted specifically to the commercialization of hybrid corn.

The founding of this company, now known as Pioneer Hi-Bred International, was the first step in a set of successive changes in the structure of the private seed industry that would have a profound effect on the nature and importance of intellectual property restrictions in the plant area.

This new plant type affected the structure of the seed industry in several important ways. First, hybrid corn revitalized a stagnant seed industry and made corn—rather than vegetables, forages, and flowers—the principal product line of the largest and most dynamic companies. Second, the commercial success of hybrid corn—its yield premium, reproductive instability, and the rapidity with which farmers adopted the new hybrid varieties from the mid-1930s through the early 1940s—stimulated continual efforts to hybridize other major field crops. Table 6.1 provides a summary of progress over time in developing hybridization systems for major U.S. crops. This list is both impressive and underwhelming; plant breeders have slowly expanded the range of crops that can be bred through hybridization techniques so as to include most of the major field crops in the United States. But it has taken a very long time to achieve hybridization for many crops—especially in the case of wheat, one of the two major U.S. field crops.

A third change set forth by the success of hybridization was an intensification of industry efforts to persuade public breeders to refrain from developing and releasing corn hybrids and finished varieties of other crops that were perceived to be privately profitable. This pressure achieved results in the early 1950s when the last corn hybrids were released by public breeders. Withdrawal from public varietal release in other crops followed shortly thereafter and is a process that has continued.

Table 6.1. Selected Crops in Which Hybrid Seed Is Currently Available

Crop	Date hybrid seed available	Hybridization system	Percent of acreage planted to hybrids, 1980
Corn	1926	Cytoplasmic male sterility (CMS)/hand emasculation	99
Sugar beet	1945	CMS	95
Sorghum	1956	CMS	95
Spinach	1956	Dioecy	80
Sunflower	?	CMS	80
Broccoli	?	Self incompatibility	62
Onion	1944	CMS	60
Summer squash	?	Chemical sterilant	58
Cucumber	1961	Gynoecy	41
Cabbage	?	Self incompatibility	27
Carrot	1969	CMS	5
Cauliflower	?	Self incompatibility	4
Pepper	?	Hand emasculation	?
Tomato	1950	Hand emasculation	?
Barley	1970	Genic male sterility	Negligible
Wheat	1974	CMS/chemical sterilant	Negligible

SOURCE: Kloppenburg (1985,203).

A fourth and final change stimulated by commercially viable corn hybrids and the rejuvenation of the seed industry was for that industry in the late 1920s to renew lobbying efforts, initiated about three decades earlier, to establish the principle and practice of proprietary protection of plant varieties. This culminated in the Plant Patent Act of 1930. Industry regarded the act as inadequate but accepted it as a useful first step because it established the principle that the private breeder should have rights to plant breeding inventions.

Thus, after the development of hybrid corn, the U.S. seed industry proceeded to develop along the following lines: 1) dominance of hybrid corn in product lines; 2) pursuit of hybridization of major field crops other than corn (which would not only compel farmers to enter the market each year to buy seeds, but would also obviate the lack of legal protection of sexually reproduced crop varieties); and 3) patenting of varieties of asexually reproduced cultivars under the Plant Patent Act of 1930. The industry continued to have a relatively dispersed ownership structure, but by the early 1960s there were roughly ten major U.S. seed companies selling field crop seeds in national (or international) markets, with hundreds more engaged in selling field crop seeds in regional markets or in selling seeds for minor vegetable and flower species.

Ratification by several European countries of the Paris Convention for the Protection of New Plant Varieties in 1960 was a further harbinger of changes in the legal framework of plant breeding in the United States. ASTA soon began to explore the possibilities for comparable legislation in the United States. Private seed companies pushed for plant breeders' rights legislation on the grounds that it would stimulate and increase the level of private investment in plant breeding and thereby benefit the nation.[7] It was also recognized that legislation protecting plant breeders' rights, if enacted, would give private firms more leverage over public breeding by strengthening their arguments that public varietal activities duplicated private efforts.

The Plant Variety Protection Act became law on 24 December 1970. It is unclear whether this law has had the intended impact of stimulating private investment in plant breeding. The number of varieties released by private companies has increased substantially, but these appear to consist largely of products of cosmetic breeding designed to be of a product differentiation nature. The rising trend in private research investment in plant breeding since passage of the the 1970 act has had the same slope as that for the decade prior to 1970 (Kloppenburg 1985). The law appears not to have led to any increase in private plant breeding investment beyond that which would have been expected from historical trends. The Plant Variety Protection Act has been more effective in protecting marketing investments than in stimulating and protecting breeding investments (Schmid 1985).

But the most important impact of this protective legislation may have had nothing to do with the stimulation of private investments. Rather, the perception that the law would increase the profitability of seed companies was an important factor in galvanizing a massive acquisition-and-merger movement involving many U.S. seed firms. The PVPA was not the only factor that contributed to the perception of long-term profitability of the seed industry. The early 1970s witnessed a tightening of world grain markets and higher commodity prices, which stimulated investor interest in all branches of the agricultural industry, including the farm-input sector. Nonetheless, the result of the merger-and-acquisition movement of the 1970s was the incorporation of many major U.S. seed companies as subsidiaries of large multinational companies, many of which have agricultural interests. Table 6.2 shows that fourteen major multinational firms—several of them from European countries—currently have as subsidiaries over seventy formerly independent U.S. seed companies. Of the ten largest U.S. seed companies in the 1960s, eight have become subsidiaries of large multinationals, and another (the second largest company,

Table 6.2. Selected American Seed Companies by Parent Firm

Parent Firm	Seed Companies
ARCO	Dessert Seed Company
	Castle Seed Company
Diamond Shamrock	Golden Acres Hybrid Seed
Cargill	ACCO
	Dorman
	PAG
	Payment Farms
	Tomco Genetic Giant
Celanese	Celpril, Inc.
	Moran Seeds
	Joseph Harris Seed Company
	Niagara Farm Seeds
Ciba-Geigy	Columbiana Farm Seeds
	Funk Seeds International
	Germain's
	Hoffman
	Louisiana Seed Company
	Peterson-Biddick
	Shissler
	Swanson Farms
	Ring Around Products
Lubrizol	Agricultural Laboratories
	Arkansas Valley Seed
	Jacques Seeds
	Keystone Seed Company
	R. C. Young
	Gro-Agri
	McCurdy Seed
	Seed Research Associates
	Sun Seeds
	Taylor-Evans Seed Company
	V. R. Seed
	Colorado Seed
Monsanto	Jacob Hartz Seed Company
	Dekalb Hybrid Wheat
	Hybritech Seed International
Occidental Petroleum	Excel Seeds
	East Texas Seed Company
	West Texas Seed Company
	Missouri Seeds
	Moss Seed Company
	Payne Bros. Seed Company
	Ring Around Products
	Stull Seeds
Pfizer	Warwick Seeds
	Clemens Seed Farms
	Dekalb AgResearch (joint venture)
	Jordan Wholesale Company
	Ramsey Seed
	Trojan Seed Company
Sandoz	Woodside Seed Growers
	Gallatin Valley Seed Company
	Ladner Beta
	McNair Seeds
	Northrup King
	National N-K
	Pride Seeds
	Rogers Bros. Seed Company
Shell Oil Company	Rudy Patrick
	Tekseed Hybrids
	Agripro, Inc.
	H. P. Hybrids
	Nickerson Seed Company
	North American Plant Breeders
	Sokota Hybrid Producers Assn.
	Ferry Morse (Farm Seed Div.)
Stauffer	Prairie Valley Seed Company
	Blaney Farms
	Stauffer Seeds
Upjohn	O' Gold
	Asgrow Seed Company
	Associated Seeds
	Farmers Hybrid Seed Company

SOURCE: Kloppenburg (1985,242).

DeKalb) has entered into a joint-venture relationship with a major multinational (Pfizer). Pioneer, the largest and most profitable company in the industry, has remained independent and become a multinational in its own right.

The significance of the acquisition of seed companies by large multinational agro-input firms lies less in increased profitability and monopoly power (which have generally not been realized) than in the potential synergies in R and D and marketing that were made possible by the rise of commercial biotechnology in the late 1970s and early 1980s. The centrality of varietal development and seed marketing to the multinational agrochemical companies was obvious; the seed was the logical commercial vector for most plant biotechnology inventions.[8] The significance of the acquisition of seed companies by multinational agrochemical companies thus lies primarily in the articulation of the seed business with the dynamism of the biotechnology investment boom of recent years.

The articulation of the seed industry and commercial biotechnology within large agrochemical multinationals has two implications that will be of major significance to the conduct of plant breeding. First, the growing attention being paid to plant biotechnology has served to intensify longstanding pressures on state agricultural experiment stations and the Agricultural Research Service to emphasize basic research and reduce applied R and D that leads to products competitive with private industry (National Academy of Sciences 1984). Such pressure on the public research system is, as noted earlier, hardly new; in the early 1970s, the so-called Pound Report chastised the public research system for its low-quality, excessively applied, and commodity-oriented research, and it recommended a greatly increased emphasis on basic or fundamental research (National Research Council 1972). These criticisms were largely ignored by most experiment stations and the research service until the early 1980s when seed companies and their multinational parent firms began to realize the commercial potentials of biotechnology. The private sector not only wished to diminish the public-sector varietal release program as before, but, more important, had a new need for more fundamental research into plant molecular biology, cell biology, genetics, and biochemistry.

These new private-sector interests, combined with stagnation of public funding of agricultural research, led to a more compelling political millieu for the restructuring of public agricultural research. The culmination of these forces was the publication of the so-called Winrock Report, which was a summary of the deliberations and recommendations of a small group of corporate, land-grant university, federal government, and foundation representatives who were convened under the auspices of the Rockefeller Foundation and the Office of Science and Technology Policy of the White House (Rockefeller Foundation 1982). The Winrock Report, like its Pound Report predecessor, stressed the excessively applied nature of state agricultural experiment station research and the need to emphasize "more fundamental research into plant and animal biology." The Winrock Report, however, went beyond the Pound Report in several respects. Most notably, the Winrock Report gave far more emphasis to the need for public researchers to devote their research to the technical needs of the private sector and stressed the important role that the public agricultural research system should play in assisting the United States in international technological competition (Kenney and Kloppenburg 1983; Buttel et al. 1984a; Buttel 1986).

Lured by the carrot of increased federal and private funding of biotechnology and fearful of the stick of more scrutiny of the public agricultural research budget, the experiment station and research service responses to the Winrock Report have been rapid by any standard. The resources of the land-grant university and agricultural experiment station system are being heavily diverted at the margin into biotechnology. Land-grant university plant breeding programs, which represent the epitome of the highly applied research that is now out of favor, have been at the heart of recent shifts of land-grant university resources toward biotechnology. Plant breeding faculties are now hiring staff members who are molecular geneticists and tissue culture specialists rather than merely wheat, tomato, or sorghum breeders. The groundwork has thus been laid for further movement along the established trajectory of the division of labor between public and private agricultural research. The public research sector is increasingly devoting its resources to more fundamental research to be transferred to industry, and is de-emphasizing highly applied research, especially that which leads to the development of products. The state agricultural experiment stations are also depending more heavily on industrial funding of research (Kloppenburg 1985; Buttel et al. 1986). The private sector is increasingly utilizing a far more fundamental level of biological knowledge, and several agro-input firms have made major commitments to the funding of agricultural experiment station biotechnology research.

A second major consequence of the union of private plant breeding and biotechnology in the late 1970s and early 1980s has been the increasingly international orientation of the seed industry. In part, the search for markets in developing countries has been stimulated by the increased competition in the industry caused by the increased investment resources made available from many of the multinational parent firms. But agro-input and seed company interests in penetrating Third World markets have other, equally important, technical and socioeconomic bases. The past few years have seen major advances in the technology for hybridizing rice and wheat, which along with corn constitute the three most important crops in the world. Hybridization of rice and wheat would make it possible for seed companies to develop long-term markets for improved varieties in low-income countries. Also, the rise of commercial biotechnology has given agro-input and seed companies an expanded range of tools for crop improvement; several biotechnologies (tissue culture methods for developing stress tolerance, for example) are especially well suited for plant improvement for developing countries. Finally, the U.S. Supreme Court in the Diamond v. Chakrabarty decision has solidified the legal groundwork for proprietary protection of plant breeding inventions to be marketed outside of the United States (Barton 1984).

In sum, the U.S. seed industry is a distinctly different entity from that which prevailed in the late 1960s. It is more concentrated, more competitive, more technologically sophisticated, has a more favorable division of labor with the public research system, and is more internationally oriented. The industry is also more firmly anchored in such intellectual property restrictions as the Plant Variety Protection Act and patents at the same time that its emphasis has continued to revolve around reproductively unstable hybrid varieties.

Propriety Protection of Plant Varieties and Plant Parts

Intellectual property (as opposed to such material property as land or personal items) includes information, products, and processes that are protected in two major ways: through statutory grants such as patents, trademarks, and copyrights; or by keeping the subject matter a trade secret (Williams 1984). The first encompasses three plant-related statutes; trade secrecy will also be discussed in the context of plant breeding and biotechnology.

The Plant Patent Act of 1930

The Plant Patent Act (35 U.S.C. Secs. 161 et seq.) provides protection for varieties of plants that can be reproduced asexually by mechanisms other than seeds—that is, by cuttings, bulbs, and so on. Although they reproduce asexually, some tuber-propagated plants (specifically, Irish potatoes and Jerusalem artichokes) were excluded from coverage, as were plants found in an uncultivated state and bacteria. The rationales for restricting protection to asexually reproduced plants were based on the belief that new plant varieties could not be reproduced reliably by seed. Some tubers were excluded because in tuber-propagated plants the propagating and edible portions of the plant are the same. Neagley et al. (1983) and others have questioned the relevance of this latter provision to protectability and have suggested that political considerations were also involved. Despite periodic efforts to remove the tuber-propagated exclusion, no implementing legislation has been passed.

According to the 1930 act (subsequently incorporated into the Patent Act in 1952), the requirements for patenting are that the variety be *distinct* (the variety must have characteristics that are clearly distinguishable from those of existing varieties) and *new* (the variety has not previously existed). An important prerequisite is that the plant must have actually been asexually reproduced before the application for the patent is considered. Requirements of patentability for utility patents that also apply to plant patents include *novelty* (the variety is "distinct and new"), *utility* (the variety serves some specific purpose), and *nonobviousness* (the variety is sufficiently different from previous varieties so as not to be obvious at the time the invention was made to someone having ordinary skill in the art).

In practice, however, novelty constitutes nonobviousness for plant patents, rendering the plant patent system closer to a registration than an examination system. One difference between plant patent requirements and those of industrial or utility patents is that a full description (the disclosure or enabling requirement) is not necessary to obtain a patent under the Plant Patent Act of 1930; for general utility patents it was mandated that the patented object be described completely enough "to enable any person skilled in the art . . . to make and use the same" (35 U.S.C. sec. 112), while for plant patents the disclosure requirement was modified to be "as complete as is reasonably possible" (35 U.S.C. secs. 161–162). Another difference is that while the Patent Act permits multiple claims of one embodiment of the invention, plant patents are granted only for the entire plant; thus, only a single claim is required and permitted.

The Plant Patent Act of 1930 bars asexual reproduction and sale of a plant covered by a patent. Enforcement against infringement is complicated by the fact that some range of equivalents is provided for in the plant patent grant, so long as the allegedly infringing plant is considered the same "variety" as the patented one.

The Plant Variety Protection Act of 1970

The Plant Variety Protection Act (7 U.S.C. secs. 2321 et seq.) extended protection to new plant varieties produced sexually by seed. The statutory framework created by the PVPA has some similarities to that of the 1930 Plant Patent Act. The PVPA, however, is administered by the U.S. Department of Agriculture instead of the Patent Office, as is the case for the Plant Patent Act. Also, under the PVPA, "certificates of protection" are issued to "breeders," while under the Plant Patent Act, "patents" are granted to "inventors" (Neagley et al. 1983).

The PVPA and its 1980 amendments provide coverage to most sexually reproduced plant varieties. Fungi, bacteria, and first-generation hybrids are excluded from protection. Hybrids were apparently excluded because of their innate biological protection and because parental lines can be maintained as trade secrets.

To be eligible for protection under the PVPA, a plant must be a novel variety—a requirement that is satisfied if there is "distinctiveness," "uniformity," and "stability." The last two conditions refer to the need for the variety to reproduce itself true to type (that is, to be homozygous). Schmid has noted that many of the traits typically used by breeders to distinguish a prospective new variety from existing ones have little to do with actual performance in the field (Schmid 1985). Indeed, the legislative history of the PVPA was one of private breeders successfully resisting any initiatives that performance of a variety be a criterion in awarding a certificate of protection (Kloppenburg 1985). Nonetheless, the breeding effort necessary to achieve homogeneity and stability needed for PVPA protection and enforcement of plant breeders' rights is, in substantial measure, cosmetic in nature—requiring a great deal of time and effort to achieve homogeneity and stability, which have virtually nothing to do with performance in the farmer's field. Moreover, the genetic homogeneity of PVPA-protected varieties may exacerbate disease and pest problems with these varieties.

The second requirement for protection under the PVPA is that the plant variety must be sexually reproduced before protection can be granted, and the third is that as complete a description of the plant as is possible must be provided, including breeding procedures and genealogy. A deposit of the seed is required for viability testing (Neagley et al. 1983).

Grounds for claiming infringement of a PVPA-protected plant (note that both the 1930 and 1970 statutes provide legal protection for plants rather than for parts of plants) are broader and more explicit than under the Plant Patent Act. The grounds for claiming infringement under the PVPA include the protections provided for under plant patents: protection against manufacture, sale, or use of a protected variety and against inducing others to perform these acts, and the prohibition of unauthorized propagation for the purpose of marketing the variety. But the provisions of the PVPA do permit farmers to use seeds of a protected variety so long as they are not engaged in marketing the variety to others. This is a significant exemption, for many farmers save and plant their own seed for such open-pollinated species as wheat, cotton, and soybeans roughly every other year. In addition, PVPA certificates protect the breeder from the importation to or exportation from the United States of the novel variety. (Importing and exporting varieties previously provided a means of circumventing U.S. plant patent protections.) Much like the Plant Patent Act, the

PVPA legislation provides for a research exemption permitting use of protected varieties in further breeding by a competing company.

Independent development of a variety is considered unlikely to occur, but if it does, it constitutes grounds for claiming infringement of a PVPA-protected variety. Interestingly, the PVPA legislation does not indicate whether the doctrine of equivalents holds—that is, whether, unlike under plant patent law, newly developed subject matter that is different but does the same thing as a protected substance may not necessarily be considered a violation of the law. The absence of an equivalence ruling means that cosmetic alterations can become the basis for receiving a certificate of protection of a "new" variety. The ability to protect cosmetically altered varieties has led Berlan and Lewontin (1983) to refer to such varieties as *varitrucs* instead of varieties because, in French, a "truc" is a trick or a dodge. Merely amending the PVPA legislation by inserting equivalence-doctrine language will not, however, be an unambiguous solution, since there is an inherent problem in defining "how different is different," especially when performance criteria cannot be brought to bear. Furthermore, it will not be easy to encourage the development of substitutes (the basis for granting proprietary rights) and yet avoid wasteful cosmetic breeding or duplication (Schmid 1985).

The Law since Diamond v. Chakrabarty

In 1980 the Supreme Court ruled in the case of Diamond v. Chakrabarty that genetically engineered bacteria constituted patentable subject material. In fact the Court ruled that patentable material includes "anything under the sun that is made by man." Following this logic, in September 1985 the Board of Patent Appeals and Interferences, in its Ex parte Hibberd decision, ruled that plants are patentable subject matter and are protectable under section 101 of the U.S. Code. This apparently includes plants presently protectable under the 1930 Plant Patent Act and under the Plant Variety Protection Act, as well as plants excluded from these acts (for example, plants propagated from tubers and first-generation hybrids). Parts of plants—roots, tubers, leaves, fruits, flowers, and seeds—can be separately protected, as can novel life-forms, chemicals, and biotechnical processes of importance, in addition to specific strains of microorganisms for conducting fermentation.

The major requirements under section 101 include novelty, utility (which includes amusement or aesthetic value), and nonobviousness. This last criterion is subject to varying interpretations. Current practice is for specialists in the field of the claim to judge how different the subject matter is from prior art; the extent to which, if at all, a significant improvement has occurred; or if there is some minimum distance in terms of the differences between the variety sought for a patent and other known varieties.

To ensure that the doctrine of equivalence does not impair efforts, for example, to patent a gene, it is likely that inventors will seek to produce matter that has a broad "patent space" (Schmid 1985). The Office of Technology Assessment (1984) has suggested the basis for this preference: "The more unpredictable the subject matter, the smaller the scope of equivalents, whereas the more pioneering the invention, the broader the scope of equivalents. Biological inventions typically involve highly unpredictable phenomena; thus, claims are likely to be narrowly interpreted" (p. 398).

Another requirement for protection under section 101 is that one skilled in the field should be able to make and use the invention as a result of the description included in the patent application. A deposit of plant material is necessary where written description is not so enabling. Under the PVPA, a description is required to be as complete as possible, but this description need not be so enabling as it must under section 101. Under section 101, however, there is no need to demonstrate derivation as has typically been the case for plant patents. The deposit procedure facilitates compliance with the enabling requirement of section 101 and provides the basis for granting general patents on microorganisms.

First-generation hybrids, which cannot be protected under the PVPA, are patentable under section 101 of the General Patent Law. Because of the nature of hybrids—heterosis (hybrid vigor) and the genetic heterogeneity between the parents and progeny—it is questionable whether deposits of the parent stock, the real key to the genetic formula, should be deposited instead of the hybrid product. An inventor can meet patent requirements by depositing the hybrid seed but can keep the parent stock a trade secret.

While only single claims and a single variety are permissible in, respectively, a plant patent and a PVPA certificate, more than one claim is allowed under section 101. Thus, it is technically possible to file simultaneous patents on a plant, a part of the plant (such as a root, leaf, or genetic component), more than one crop species (provided material is available for deposit), and on a process as well as the products of the process (Neagley et al. 1983). Grounds for infringement under section 101 are similar to those for plant patents—namely, unauthorized manufacture, sale, or use of the protected material. The doctrine of equivalence applies such that independent invention (whether intentional or not) constitutes infringement. In addition, since plant parts are protectable material, and because there is no exemption for farmers to save their seeds, section 101 patenting provides a broader range of protection than do the Plant Patent Act and the PVPA. These advantages—along with the fact that section 101 protection can be achieved more cheaply than protection under the PVPA—

make section 101 patenting the preferred means of protecting plant-related inventions by private companies in the United States (Barton 1984; Adler 1984; Office of Technology Assessment 1981).[9]

As with the PVPA, research on a patented seed product is generally seen as a fair use. The major difference between section 101 plant patenting and the PVPA, however, lies in the fact that if patent protection of a component (that is, a DNA sequence) of a patented seed is extended to a claim over a range of species, a license will be required to use that component in another variety. This feature contributes further to the greater security of section 101 plant patenting over plant variety protection.

Trade Secret Protection

Any inventor has the prerogative of keeping products or processes secret. This right is maintained through affirmative acts on the part of the trade secret holder, such as getting employees to sign documents attesting to their knowledge that the information is secret and their concurrence not to disclose the content of the secret in an unauthorized way. State courts enforce trade secret rights by serving injunctions against unauthorized disclosure or use of protected information and by compensating secret holders (Saliwanchik 1982).

Trade secrets are desirable where an invention is clearly nonpatentable, the patent right is unenforceable, or the expense of enforcing patent rights would be uneconomical. A major advantage of protecting important discoveries or inventions through the trade secrecy approach is that competitors are not informed of the nature of the discovery, as would be the case through the descriptions required under plant patent, the PVPA, and general patent legislation. This consideration may be particularly important if there is a time lag between submitting an application and receiving approval of a patent or if the pace of technological discovery is rapid (Adler 1984). Private companies wishing to protect their market for a new product or process may choose to rely on trade secrets even for patentable inventions in order to avoid disclosure; rapid market penetration by the new product may serve the firm's commercial objectives without the risk of providing technical information to competitors through patent documents.

The advantages of avoiding disclosure must, however, be weighed against the risk of theft or unauthorized trade secret leaks and against the lack of protection against a competitor who examines a finished product and who is able to reverse engineer to the trade secret (Office of Technology Assessment 1984). Those relying on trade secrecy also run the risk of having to litigate should a competitor make the same discovery and apply for a patent (Barton 1984). This is particularly risky when many researchers are working on similar problems using nearly identical methods (Fox 1984).

Trade secrets are not so prestigious as patents for a company that can receive benefits (for example, higher stock or asset values) by being able to measure symbolically its technical ability by the number of patents it holds (Fox 1984). Further, the trade secrecy approach denies scientists the opportunity to publish results, whereas patent protection allows for such rewards and personal recognition. This discrepancy can create conflicts when private companies and university faculty engage in extramurally funded, collaborative research (David 1982).

Before the Chakrabarty decision, private firms either relied on trade secrets alone or used a combination of patents on microbiological processes and products with trade secret protection of microorganisms. There are four major advantages to forgoing the trade secret approach by obtaining a Section 101 patent.

First, where the microorganism is the product (such as with Chakrabarty's oil-consuming bacteria), patenting the organism is the best means of protecting the invention, given the above-mentioned advantages and limitations of patenting and trade secrecy.

Second, patenting is the most secure means of reaping economic returns from licensing the use of a microorganism, gene, or process to other companies.

Third, patenting inhibits a competitor's ability to obtain the microorganisms and expands the grounds for demonstrating infringement.

And fourth, patenting protects the inventor's rights to uses and products of the life-form that were not discovered by the inventor; royalties have to be paid to the inventor for any use of a life-form for commercial purposes (Office of Technology Assessment 1981).

Patenting, Biotechnology, and Incentives

The ability to patent new biotechnologies is encouraging private industry to allocate increasing levels of R and D resources to the development and commercialization of such technologies (Adler 1984). Expanding opportunities for proprietary protection of novel life-forms and the processes to create them are also beginning to revolutionize private-sector plant breeding, much as this change is influencing other life-science industries, such as pharmaceuticals, chemicals, and energy.

While intellectual property restrictions play an increasingly important role in American plant breeding, changes in legal institutions cannot account for this entire revolution. Indeed, proprietary protection of plant inventions has assumed greater importance because of certain technical advances and because of the restructuring of the seed industry.

As noted earlier, two technical advances have increased the incentive for seed companies to give greater attention to intellectual property restrictions on their varietal inventions. First, advances in plant molecular biology, cell biology, and biochemistry have provided private (and public) breeders a number of potentially powerful tools to increase the speed and effectiveness of plant genetic manipulation. In particular, these tools give private breeders the ability to transcend the often cosmetic breeding that has characterized private seed companies' efforts to compete on the basis of product differentiation. Second, there have recently been technical advances in hybridizing wheat and rice—the two major world crops that private companies had thus far been only partially successful in penetrating. Private firms in the developed countries are currently testing several hybridization systems for wheat and rice. Most of these systems revolve around using gametocides that induce male sterility (Barton 1984). It is possible that chemically induced male sterility of this sort may not be patentable, for the gametocides are not novel chemicals, but biotechnological procedures may make possible the achievement of cytoplasmic male sterility, which presumably will be patentable. For the present time, private breeders of wheat and rice hybrids will rely heavily on trade secrecy (with respect to both gametocides and the genetic composition of hybrids). As private breeders are able to incorporate novel genes in their varieties, these inventions will tend to be protected under the General Patent Law.

Earlier in this chapter we referred to several changes in the structure of the seed industry that bear on intellectual property issues. Two of these changes are important here: the greater availability of investment capital as a result of the acquisition of seed companies by agrochemical multinationals, and the heightened degree of competition in the industry. The first factor provides the financial resources to utilize new techniques (especially biotechnologies) in plant breeding, to comply with the more stringent patenting requirements of section 101 (that is, the increased need to demonstrate distance from prior art), and to protect these inventions through various intellectual property restrictions. The second factor, in effect, compels private seed companies to do so—that is, to make larger investments, assume more risks, and to take added precautions vis-à-vis competitors in order to maintain existing markets and penetrate new ones.

It is crucial to recognize that seed companies are now often subsidiaries of agrochemical companies. The multinational parents of seed companies have large fertilizer, herbicide, insecticide, and fungicide product lines that generally are far more important in terms of total revenue and profit than are seeds. Among the major potentials of plant biotechnology are the abilities to develop varieties with nitrogen-fixing capability and to develop bacterial pesticides that are safer and cheaper than synthetic organic chemicals. Such new products would be fertilizer and chemical displacing. Accordingly, many agrochemical-based seed company subsidiaries might be hesitant to emphasize plant breeding goals that would threaten fertilizer and pesticide product lines.

One can thus explain why a substantial amount of plant research in private firms has been aimed at developing various types of seed-chemical packages that reinforce rather than threaten sales of agricultural chemicals. Probably the single most important area of recent research in plant biotechnology is that of herbicide tolerance. Herbicide-tolerant DNA sequences are patentable under the General Patent Law; Calgene, for example, has recently received a patent for a novel DNA sequence that enhances herbicide tolerance in tomatoes, tobacco, soybeans, and cotton. Agrochemical-based seed companies are utilizing tissue culture techniques to develop corn varieties that are tolerant of herbicides produced by those companies.[10] They hope to sell seeds and herbicides to farmers as a package (although it is widely acknowledged that there are formidable biological and commercial barriers to the packaging of seeds and chemicals). The new synthetic seeds, for example, consist of pregerminated embryos encapsulated by a gel containing plant-protection chemicals. Such research may replace attempts to use biotechnologies and conventional breeding techniques to achieve pest-resistant varieties that would reduce the dependence of farmers on chemical pesticides.

Barton has noted a comparable dilemma with regard to nitrogen fixation in the cereal grains (Barton 1984). At the beginning of the biotechnology investment boom, one of the most highly touted areas of application was the potential for achieving nitrogen fixation in such cereals as corn, wheat, and rice. Many of the private companies that pursued biotechnology at an early stage had nitrogen-fixing research programs. But it appears that many of these programs have been abandoned—in part, because nitrogen fixation is technically complex (and expensive) and perhaps also because of the conflict with fertilizer product lines of agrochemical investors. This strategy, however, involves the risk that competitors will find an inexpensive, novel nitrogen-fixing mechanism that could become commercially valuable (Barton 1984).

Private firms have also become interested in obtaining patents on new biotechnological processes that would have broad use in the industry. Private-sector patenting activity in the plant area has been more lively in process patents than in such product patents as novel genes, DNA sequences, or whole plant varieties following the Ex parte Hibberd decision. If a firm can develop and patent a process with broad applicability in the industry, the firm can deny the process to competitors or receive a royalty for use of the process. Among the more important process patents in the seed industry is Agrigenetics' (a subsidiary of Lubrizol) hybrid seed production process. ARCO, Calgene, DNA Plant

Technology, International Plant Research Institute, and Phytogen have also received or have major patents pending on plant processes (Kloppenburg 1985).

Public breeders have long loathed process patents, since the latter affect the ability of public researchers to use these processes and to serve the seed industry as a whole. Agrigenetics' hybrid seed production patent, for example, proved to be quite controversial, because many public breeders felt that portions of the Agrigenetics process involved techniques that had previously been used by plant breeders (see Appendix, below). Nonetheless, indications are that private companies are placing major emphasis on acquiring process patents in order to increase their leverage over competitors and over public breeding programs (Kloppenburg 1985).

Process patents with broad "patent space" involve, however, a high probability that the patent will not be respected or that competitors will challenge the patent in the courts. There is, for example, consensus that the Stanford and University of California patent application regarding the Cohen-Boyer process (see Boonin this volume) is too broad—that it will be difficult for Stanford and the University of California to protect their patent rights in such an important and widely used process (Lewin 1983).

A holder of a product or process patent can profit from an invention in two major ways: through the marketing of finished products and through royalty income from licensing the patented product or process to other entities. Realizing returns from patented inventions, especially in plant breeding, involves vertical integration in which a single firm conducts research, produces seed, and markets the seed. By marketing finished products and thereby realizing returns from patents, a firm can take advantage of economies of scale, integration of management, and longer-range planning and can expand its market share (Barton 1984). Thus, intellectual property restrictions in the plant sector may increase the already high level of vertical integration among seed companies that are owned by agrochemical multinationals.

Many of the firms doing plant biotechnology research, however, are start-up companies that lack seed-marketing outlets. Although there are not high barriers to entry in the seed production and marketing industries, most viable national or international seed companies in Western countries have already been purchased (Table 6.2), and the cost of establishing a new seed production and marketing company would be prohibitive for virtually all small start-up companies. Thus, the biotechnology start-up companies have little choice but to use their patents to earn royalty income or, in some cases, to acquire bargaining chips with larger companies for use of their patented processes or products. The material in the Appendix highlights another strategy vis-à-vis patents—to bolster asset values—that may have controversial implications for the industry.

Some small plant biotechnology companies are also considering the unusual step of releasing their patented technologies to other small firms without royalty charges in order to inhibit the oligopolization of the seed industry (Schmid 1985).

One of the most crucial aspects of the incentive structure for private plant breeding firms concerns which types of crops will be given emphasis. Seed companies in the United States are, not surprisingly, devoting increasing emphasis to such crops as corn, wheat, sorghum, soybeans, cotton, and tomatoes, which have large markets, while decreasing their emphasis on minor crops (Busch et al. 1984). Given the increased integration between seed companies and universities, land-grant university researchers are beginning to follow suit in increasing their emphasis on major field crops, vegetables, and fruits.

The market-driven nature of plant biotechnology aimed at Third World markets is equally apparent—and may perhaps have more dramatic consequences. The bulk of private-sector interest in Third World seed markets has been focused on wheat and rice—the two major green revolution crops, as well as the two Third World field crops that have exhibited the highest degree of commercialization—along with maize. The interest in wheat and rice has been made possible, as noted earlier, through a combination of advances in hybridization, internationalization of the seed industry, and the availability of new biotechnology techniques and proprietary protections. It is interesting to note that the third most important field crop in the Third World—maize—has received far less emphasis despite the fact that corn clearly remains the most important product line of First World seed companies and that there has been a large amount of research on maize in North America and Western Europe. The deemphasis of maize can be explained largely by the fact that Third World maize producers do not tend to be highly commercial; they are usually peasant smallholders who grow maize in association with other crops and largely for home consumption. Maize producers thus tend not to constitute attractive markets for multinational seed companies. This leads to limited resources' being devoted to maize breeding for Third World markets.

One of the most controversial aspects of the history of plant breeding has been the debate over the benefits and costs of the green revolution (Griffen 1974; Wortman and Cummings 1978). Among the major criticisms of the green revolution has been the claim that the adoption of high-yielding green revolution varieties of wheat and rice was socially and geographically uneven. Critics contend that adoption of these varieties was often more beneficial to larger farmers than to peasant smallholders, and that in most Third World countries, the green revolution exacerbated rural and agricultural inequality. Critics have also emphasized the fact that high-yielding green revolution varieties were suitable only for highly productive agro-ecological zones and that less-favored zones were further marginalized as a result of the spread of green revolution varieties.

To the degree that the green revolution was socially and geographically unequal in the distribution of costs and benefits, the private-sector-led "biorevolution" is likely to be even more so (Buttel et al. 1984b; Buttel et al. 1985). The biorevolution will not only build on the inequalities of land and on the degree of commercialization that were reinforced by the green revolution in wheat and rice, but also is likely to exacerbate these inequalities.[11]

The green revolution progressed so rapidly in part because the high yielding varieties of wheat and rice were all reproductively stable nonhybrids. Farmers could save their own seed, which placed a ceiling on seed costs and reduced farmers' outlay for purchasing planting material. The biorevolution, which will revolve around the development and marketing of reproductively unstable hybrids and will embody proprietary genetic material and processes, will lead more Third World farmers to enter the market each year for seeds that (by comparison with green revolution standards) will probably be relatively expensive. Given persisting inequalities of access to credit, not all Third World peasants will be able to afford these new seeds. Those who cannot will be forced to grow other crops or to leave the agricultural sector.

Likewise, whole countries—those with small numbers of commercial wheat and rice producers—will have limited access to the products that result from biotechnological innovation in plant breeding; private companies driven by market constraints will probably tend to de-emphasize small countries with less fertile lands just as much as they will de-emphasize minor crops in developed-country markets.

Discussion

This chapter has been concerned with analyzing the structure of the domestic and global seed industry and with discussing some of the likely effects of intellectual property restrictions on innovation in this industry. We would like to emphasize several points in conclusion. First, although biotechnology in combination with new intellectual property institutions has created an unprecedented situation in the plant breeding industry, intellectual property issues have long played a major role in the development of this industry. For many years, private breeders have struggled to establish proprietary exclusivity to their breeding inventions. The rise of commercial biotechnology in the late 1970s, the 1980 Supreme Court decision in the Diamond v. Chakrabarty case, and the 1985 Ex parte Hibberd ruling are but the latest chapters in this long drama.

Second, the importance of intellectual property restrictions to the contemporary plant breeding industry has not been caused by legal changes alone. New technologies (hybridization of wheat and rice and the availability of new biotechnology tools for genetic manipulation) and structural changes in the seed industry (the acquisition of seed companies by large agrochemical multinationals) have made intellectual property issues more important than in the 1960s.

Finally, the productivity increases made possible through proprietary plant biotechnologies must be balanced against possible adverse impacts on several groups in the United States and abroad. In the United States, one can expect that these new arrangements may harm several portions of the public breeding program—for example, breeders who wish to release finished varieties or who focus on minor crops. These arrangements may also provide less benefit to producers of minor crops than to corn, wheat, sorghum, rice, cotton, or tomato producers. In Third World contexts, the major beneficiaries of the private-sector-led biorevolution are likely to be large wheat and rice farmers in large countries with highly productive agricultural resources. Producers of other crops in smaller countries with unproductive lands will benefit less or may even be hurt in relative terms.

Appendix: Agrigenetics and Agricultural Genetics

Jack Kloppenburg

On 27 April 1982, U.S. Patent No. 4,326,358 was issued. Assigned by the inventors to Agrigenetics Research Associates Limited, the patent's simple descriptive title—"Hybrids"—gave little indication of the importance of the claims made in the body of the text. Yet the descriptive title was most appropriate in its breadth, for Patent No. 4,326,358 had the potential to establish property rights over an extremely broad area. The patent made fourteen separate claims, but in its essentials it gave Agrigenetics Research Associates Limited the rights to the process of using clonally propagated parental lines to develop new hybrid plant varieties.

The application of the new biotechnology to plant improvement has received a great deal of attention. One of the most promising fields of research is the use of tissue culture to reproduce genetically identical copies of a plant line (cloning). If the patent can hold up against challenge in court, it could give Agrigenetics Research Associates considerable control over one of the cutting edges of the field of plant breeding. Anyone using clonally propagated breeding lines would have to come to Agrigenetics for licensing of the technology. Patent No. 4,326,358 is potentially as critical to plant breeding as the patent of the Cohen-Boyer process is to the field genetic engineering. Ownership of so fundamental a patent could thus confer a considerable commercial advantage on Agrigenetics. Trumpeting what was called a "major technological breakthrough," Agrigenetics, in press releases reported in the June 1982 *Seed World* (120:18) and the July 1982 *Seedsmen's Digest* (33:9), announced that it would commence licensing negotiations with interested parties immediately and would "aggressively defend its patent position, energetically use discovery procedures, and actively pursue any entity infringing on its proprietary patent rights." Both the seed industry and the plant science community were put on notice that in the brave new world of biology, science is the handmaiden of business, and that at the core of business are property rights.

A response was not long in coming; it was not, however, the response Agrigenetics wanted to hear (though, as shall be seen, it was one they must have expected). In a letter published in the 26 August 1982 issue of *Nature*, N. L. Innes, a prominent British public plant breeder and chairman of the British Association of Plant Breeders, accused Agrigenetics of arrogating to itself "rights over techniques that have been part of the stock-in-trade of plant breeders for some considerable time and have already been used commercially" (298:786). Innes went on to assert that the patented principles and techniques were known and practiced before the patent was filed, that the particular combination of techniques detailed in the patent had been recognized, and that the entire approach was obvious to anyone with ordinary skill in plant breeding. He concluded by bemoaning what he regarded as a bald-faced attempt to "restrict the use of techniques and combinations of techniques that are common currency among plant breeders worldwide." Essentially, Innes declared that the emperor had no clothes, that Agrigenetics' "breakthrough" was established practice, and that, by implication, the patent should never have been issued. In a very unusual action, at its meeting of October 1982 the board of the European Association for Research on Plant Breeding reviewed the issue and concluded that "none of the techniques in the patent [is] new; some of them have actually been applied for ages" (*Euphytica* 33:2).

In the United States sentiment among both publicly and privately employed plant breeders seems consistent. There is virtually unanimous agreement in the plant science community that the Agrigenetics patent is a "questionable novum." Breeders cite prior art in a variety of crops ranging from sugar beets to cabbage. One public forage breeder has observed of the patent: "It's as ridiculous as anything I have ever heard. My first reaction was one of real anger that anybody would even try this. I mean, the use of clonal lines and crosses—we've got cultivars that are in production now that are developed exactly like they state in the patent, and they have been in production for twenty years." And a vegetable breeder at a major seed company concurred: "Every major brassica producer has used plant tissue culture since the 1970s. . . . No one ever thought to patent what was common knowledge." The then-chairman of Agrigenetics defended the patent in an interview in *Nature*, saying that "any given detail or sequence may seem obvious, but the way they're put together may be original" (298:782). He insisted that Agrigenetics would continue to attempt to license the patent for use at "fair and reasonable" terms.

To date, although the company has made an effort to license the technology, no one has taken Agrigenetics up on its "fair and reasonable" terms, and the patent remains unlicensed. In an article in the *American Vegetable Grower*, Asgrow Seed Company (a subsidiary of Upjohn) made it clear that it was using the patented techniques in its breeding program. According to one Asgrow executive, the company received a letter from Agrigenetics noting that it was apparent that Asgrow was using Agrigenetics' technology and insisting that they obtain a license for its use. Asgrow responded by citing scientific literature that it felt represented prior art that would invalidate the patent in litigation and asked for Agrigenetics' response. No reply was forthcoming from the biotechnology company. Asgrow's management has reportedly told its researchers not to worry about infringements, that the patent is "not worth the paper it is printed on."

Public researchers have taken a similar tack. The patent is widely regarded as indefensible in a court of law. Public breeders, too, continue to use the patented techniques in the belief that Agrigenetics would be unable to prosecute successfully any infringement. For its part, although it must be aware that many individual breeders—both public and private—are using the techniques detailed in Patent No. 4,326,358, Agrigenetics has not moved energetically to protect its rights and defend the patent. Publicly, company executives state that Agrigenetics intends to pursue infringements, believes implicitly in the legitimacy of the patent, and has every confidence that its rights would be upheld in a court of law. However, the company has chosen not to litigate although there are known infringers. And at least one officer of the firm privately admits that he feels the patent "has had a real negative impact on [the company's] credibility and respect; it's a bad precedent."

What, then, is the significance of this episode? It might appear that the story of Patent No. 4,326,358 has little to tell. After all, no one is respecting it in practice. It may be illegitimate, but it is ignored in any case. Is not "justice" being served? Where is the problem? Yet to take such a narrow view is to miss a number of issues that are imbedded in the problem of Patent No. 4,326,358. In particular, the case is best understood not as an isolated incident but in the context of the continued development of plant breeding as a science and as a business, with the social, economic, and structural changes that such development implies. In the remainder of this case study, I will lay out a number of issues that I feel emerge from the story and have ramifications for the broad question of property rights as related to the conduct of science.

It appears that only Agrigenetics and the Patent Office believe that the patent is legitimate. Plant breeders can cite a whole series of instances of prior art. Agrigenetics, whatever it might say publicly, has implicitly confirmed the questionable nature of its patent by its failure to pursue gross and publicly recognized infringements by both public and private breeders. In defense of their critique of the patent, many breeders point to the references cited in the body of the patent. Of the ten publications cited in support of Agrigenetics' application, six were articles in the *Encyclopaedia Britannica*, two in the trade journal *American Vegetable Grower*, and one in the *Journal of the American Society for Horticultural Science*. Plant scientists do not regard these popular and semipopular articles ("primary school references," in the words of one breeder) as an adequate review of the scientific literature relevant to the allegedly new processes detailed in the patent. One question that arises is why the Patent Office approved the patent on what appears to be thin evidence. What was the role of the patent examiner, and are existing procedures adequate to ensure effective review?

Another question relates to these: If the citations to existing art are inadequate, might there have been a willful disregard of relevant literature on the part of the inventors? In other words, might the patent have been sought even though executives of Agrigenetics realized its tenuous legitimacy. This is a reasonably serious allegation, and what follows is largely speculation. Absent an admission to this effect by one of the principals, this possibility cannot be confirmed. Yet, such a scenario, as will be shown, would actually represent sound corporate strategy, and the possibility needs at least to be considered.

As yet, few actual products have resulted from the application of the new biotechnologies to plant breeding. In order to raise the financing needed to perform research, plant biotechnology companies such as Agrigenetics have had to depend as much (and often more) upon outside funding as on product sales. Such sales require publicity. The patent thus could have been pursued more for its publicity and fund-raising value than for its actual scientific legitimacy or commercial potential. In the absence of products, patents become the visible sign of a biotechnology company's vitality. Agrigenetics was seeking secondary financing, and it issued a prospectus in December 1983 in preparation for an issue of common stock (the process of going public was subsequently aborted, and Agrigenetics was purchased by Lubrizol in late 1984). In its prospectus, the company emphasized that its newly patented breeding "technology will permit additional rapid product introductions in the future."

If the patent was so tenuously based, why was there not an outcry in the

United States similar to the reaction in Europe? If the board of the European Association for Research on Plant Breeding (EUCARPIA) could flatly state that "none of the techniques in the patent [is] new," why were U.S. organizations silent? If no one believes the patent is legitimate, why has there been no challenge?

Some observers argue that there is a perceptible reluctance among U.S. companies to give biotechnology bad press. Perhaps more important, private capital is now engaged in efforts to extend the reach of property rights. The potentials of biotechnology are galvanizing shifts in what can be brought under the commodity form. Private industry is therefore reluctant to jeopardize current attempts to extend property rights to new areas. Since the Agrigenetics patent did not actually constrain its own research—it is so indefensible that serious suits from Agrigenetics are not anticipated—the corporate world has elected to avoid controversy that might undermine the legitimacy of patents in general.

Corporate influence is felt in the professional scientific societies as well. While public breeders in the United States might have felt that some institutional response similar to EUCARPIA's was warranted, private interests in the Crop Science Society of America (CSSA) and the American Society for Horticultural Science counseled caution. According to one officer of the CSSA, the society's reluctance to get involved in such controversial matters reflects the "diversity of opinion" resulting from the society's combination of public and corporate membership.

There is also the problem of the effect of Patent No. 4,326,358 on information exchange and the willingness of researchers to pursue particular lines of inquiry. All the public and private breeders I contacted have indicated that they are continuing to use the patented techniques. My sample is hardly exhaustive, however, and it includes mostly prominent investigators at well- known institutions. At the margins, the threat of litigation may have constrained research in small seed companies or small universities. The case has certainly contributed to the trend toward a narrowing of the types of information that scientists are willing to exchange, and particularly to a growing awareness of the commercial implications of the work accomplished in universities and other publicly funded institutions. As a result of the patent, one breeder in a private company commented: "I suspect the public side will be a lot more careful; it seems already obvious they are not so willing to divulge their information. They say, 'Wait until it's published.'"

In sum, it would be a mistake to regard the case of Patent No. 4,326,358 as simply an isolated incident, nothing more than a matter of hubris that, if ignored, will evaporate. The case raises basic questions as to the operation of the Patent Office, the possibility of taking out frivolous patents for strategic business purposes, the effect of corporate influence on the operation and policies of professional societies, and the changing nature of information flow among researchers.

Notes

Acknowledgments: This research was supported by the Cornell University Experiment Station. The authors would like to extend their deep gratitude to Jack Kloppenburg for this helpful comments on previous drafts of the manuscript and, in particular, for his preparation of the material that appears in the Appendix.

1. Despite the importance of cereal grain seeds in the product lines of most contemporary seed companies, it is interesting to note that the earliest seed companies in the United States sold primarily vegetable and flower seeds.

2. By the 1970s, however, private-sector agricultural research and development had grown to the point where private firms accounted for over half of agricultural R and D in the United States; see Ruttan (1982).

3. Agricultural-input firms have had such leverage over public research institutions because agribusiness has long been the major supporter of public research. Put somewhat differently, farmer organizations have not tended to be enthusiastic about agricultural research, since farmers tend to suffer lower commodity prices and profits from research that increases yields and aggregate production; see Kloppenburg and Buttel (1987).

4. This section draws heavily from Kloppenburg (1985).

5. The establishment of a satisfactory degree of research capacity in the LGUs occurred over a very long period, from the 1860s to the 1920s.

6. This was principally because vegetables, flowers, and forages are not usually allowed to grow to maturity—the stage at which seeds are produced.

7. Private seed companies were also successful in insisting that performance criteria would not be a consideration in awarding certificates of protection.

8. The biotechnology industry has a dual structure, consisting largely of (1) large chemical and pharmaceutical multinationals, and (2) small start-up companies established during the late 1970s and 1980s. There are, however, growing links between these two types of companies. Many multinational companies have made major investments in start-up firms, and start-up companies typically seek to sell or license their technologies to multinationals. The most important reasons for this integration of large and small companies are that the small companies lack products (and hence revenues and profits) and lack the scale-up and marketing expertise to take new discoveries from the laboratory to the marketplace. See Kenney (1986).

9. Most European countries, however, prohibit utility patents for seeds—seeds patented in the United States must be protected by other means in Europe and elsewhere overseas. See Lesser (1986) for an especially thorough discussion of these and related aspects of plant patenting since Ex parte Hibberd.

10. A number of start-up companies are conducting biotechnology research aimed at identifying and patenting plant genes that would be the basis for seed-chemical packages

to be marketed by large multinationals. For example, Calgene, a start-up company in Davis, California, has succeeded in identifying genes that confer tolerance to Roundup, the leading grass herbicide for corn that is marketed by Monsanto. Calgene apparently hopes to profit from this reasearch by licensing the technology to Monsanto in return for royalty income. Should a company such as Monsanto have access to genetic material conferring tolerance to one of its pesticides, it would be unlikely to restrict access to this genetic material. Broad diffusion of such a gene in advanced breeding lines provided by public breeders and in the varieties of competitors would enable Monsanto to increase pesticide sales.

11. However, we have argued elsewhere (Buttel et al. 1985) that these dislocating and differentiating effects of agricultural biotechnologies in the Third World can be mitigated through strengthening the International Agricultural Research Centers, whose efforts will bear the most fruit if they continue to emphasize, as they have done recently, the technical needs of peasant small holders.

References

Adler, R. G. 1984. "Biotechnology as an Intellectual Property." *Science* 244: 357–363.

Barton, J. H. 1984. "The Effects of the New Biotechnologies on the International Agricultural Research System." Report prepared for the United States Agency for International Development.

Berlan, J. P., and R. Lewontin. 1983. "Breeder's Rights and the Patenting of Life Forms." Unpublished manuscript, Institut National de la Recherche Agronomique, Centre d'Economie du Développement Comparé, Université d'Aix, Marseille.

Busch, L., M. Hansen, J. Burkhardt, and W. B. Lacy. 1984. "The Impact of Biotechnology on Public Agricultural Research: The Case of Plant Breeding." Paper presented at the annual meeting of the American Association for the Advancement of Science, New York, May.

Buttel, F. H. 1986. "Biotechnology and Agricultural Research Policy: Emergent Issues." In *New Directions for Agriculture and Agricultural Research*, edited by K. A. Dahlberg, pp. 312–347. Totowa, N.J.: Rowman and Allanheld.

Buttel, F. H., J. T. Cowan, M. Kenney, and J. Kloppenburg, Jr. 1984a. "Biotechnology in Agriculture: The Political Economy of Agribusiness Reorganization and Industry-University Relationships." In *Research in Rural Sociology and Development*, edited by H. K. Schwarzeller, pp. 315–348. Greenwich, Conn.: JAI Press.

Buttel, F. H., M. Kenney, J. Kloppenburg, Jr., and J. T. Cowan. 1984b. "Biotechnology in the World Agricultural System: A New Technologic Order for the New Biology?" Paper presented at the annual meeting of the American Association for the Advancement of Science, New York, May.

Buttel, F. H., M. Kenny, and J. Kloppenburg, Jr. 1985. "From Green Revolution to Biorevolution: Some Observations on the Changing Technological Bases of Economic Transformation in the Third World." *Economic Development and Cultural Change* 34: 31–55.

Buttel, F. H., M. Kenney, J. Kloppenburg, Jr., and D. Smith. 1986. "Industry-University Relationships and the Land-Grant System." *Agricultural Administration* 23: 1–35.

David, E. E., Jr. 1982. "The University-Academic Connection in Research: Corporate Purposes and Social Responsibilities." Paper presented to the New York City Bar Association, April.

Fox, J. L. 1984. "Gene Splicers Square Off in Patent Courts." *Science* 224: 584–586.

Griffen, K. 1974. *The Political Economy of Agrarian Change*. Cambridge: Harvard University Press.

Kenney, M. 1986. *Biotechnology: The University-Industry Complex*. New Haven: Yale University Press.

Kenney, M., and J. Kloppenburg, Jr. 1983. "The American Agricultural Research System: An Obsolete Structure?" *Agricultural Administration* 14: 1–10.

Kloppenburg, J., Jr. 1985. "First the Seed: A Social History of Plant Breeding and the Seed Industry in the United States." Ph.D. diss., Cornell University.

Kloppenburg, J., Jr., and F. H. Buttel. 1987. "Two Blades of Grass: The Contradictions of Agricultural Research as State Intervention." In *Research in Political Sociology*, vol. 3, edited by Richard G. Braungart, pp. 111–135. Greenwich, Conn.: JAI Press.

Lesser, W. 1986. "Patenting Seeds: What to Expect." Unpublished manuscript, Department of Agricultural Economics, Cornell University.

Lewin, T. 1983. "Genetic Engineering: To Patent Everything in Sight." *Technology Review* 86: 33–34.

National Academy of Sciences. 1984. *Genetic Engineering of Plants*. Washington: National Academy Press.

National Research Council. 1972. *Report of the Committee on Research Advisory to the United States Department of Agriculture*. Washington: National Academy Press.

Neagley, C., D. D. Jeffrey, and A. B. Diepenbrock. 1983. "Section 101 Plant Patents: Panacea or Pitfall?" Paper read at the annual meeting of the American Patent Law Association, Plant Variety Protection Committee, Crystal City, Va. October.

Office of Technology Assessment. 1981. *Impacts of Applied Genetics*. Washington: Government Printing Office.

______. 1984. *Commercial Biotechnology*. Washington: Government Printing Office.

Rockefeller Foundation. 1982. *Science for Agriculture*. New York: Rockefeller Foundation.

Ruttan, V. W. 1982. *Agricultural Research Policy*. Minneapolis: University of Minnesota Press.

Saliwanchik, R. 1982. *Legal Protection for Microbiological and Genetic Engineering Inventions*. Reading, Mass.: Addison-Wesley Publishing Company.

Schmid, A. A. 1985. "Biotechnology, Plant Variety Protection, and Changing Property Institutions in Agriculture." *North Central Journal of Agricultural Economics* 7: 129–138.

Williams, S. B., Jr. 1984. "Protection of Plant Varieties and Parts as Intellectual Property." *Science* 225: 18– 23.

Wortman, S., and R. W. Cummings, Jr. 1978. *To Feed This World*. Baltimore: Johns Hopkins University Press.

Current Controversies

PART III

It is not surprising that controversies arise over the use of intellectual property protections when there appear new areas of technology with tremendous commercial potential. The chapters in this section focus on two such technologies, especially interesting because they have led to modifications in intellectual property law, designed to accommodate their special needs. We discuss here controversies over the use of copyrights to protect computer software and over the use of patents to protect biotechnology.

Software and Copyrights

The software industry has generated controversies over intellectual property throughout its short history. It is a commonplace to say that software resembles both patentable and copyrightable subject matter, and seems to be only imperfectly protectable by either. As a consequence, the preferred form of protection has until recently been trade secrets, established through special contracts between users and producers. There have also been a number of suggestions for special laws to create a generic form of protection for just that industry. Recently, however, software innovators have embraced copyrights as their favored form of protection, in spite of a number of important reservations.

To appreciate the importance of copyrights for software, we must note that patent protections have been denied to software innovators, even though patents seem intuitively more appropriate than copyrights. The intuitive point is that software is used to perform tasks (run machines, control processes, generate data, and so on) and that processes, machines, and gadgets that perform functions are the traditional subject of patents. Still, in a number of significant decisions, the Supreme Court has consistently denied patents on software. (This point is popularly misunderstood since in Diamond v. Diehr, 450 U.S. 175 [1981] a deeply divided Court upheld a patent on a complex of machinery that included special software in one of its components. But even in that decision, both sides repeated their basic opposition to patents that primarily protect software innovations.)

Although the theoretical basis for denying patents on software is not fully relevant to the debate of this section, it is interesting as a background for the debate. The ban on software patents is an extension of a traditional rejection of patents on theoretical, mathematical methods. Although inventions inspired by discoveries in mathematics can be patented, the mathematics itself has always been seen as public domain and free to all researchers without reservation. This distinction between the machine or physical process (ownable) and the theory that underlies it (unownable) is central to all patent law. Since the most ingenious aspects of innovative software are algorithms that can be viewed as ways to carry out mathematical calculations, they are too much like theorems of mathematics to be protectable. Although this view of algorithms and innovative software has caused considerable difficulty for computer science, it has considerable merit. Nevertheless, some skeptics have argued that the real reason for denying patents is that the Patent Office would be totally unable to deal with the surge of patent applications (each of which takes considerable time to process) that would follow a favorable decision for software patents. In any event, we have arrived at a point where software researchers really cannot expect patent protection. (In 1989 the Patent Office began to soften its opposition to patents on algorithms, and thereby kindled a new debate over how one should view applied mathematics.)

We may speculate that copyrights are permitted on software because patents are not, and the courts realistically recognized the need to grant the industry some protection. Copyrights are an unintuitive way to protect innovative ways to perform processes. Copyrights protect sequences of words that appear in a program, that is, the manner in which the algorithms are coded, and not the algorithms themselves. (This is the counterpart in copyright law of the patent law prohibition on owning theoretical mathematics.) In theory, it is therefore possible for someone to read a copyrighted program, rewrite the algorithms in new code, and distribute or use the new programs without infringing the original copyright. (In practice, it is hard to distinguish copies with slight variations from new programs in different code, and rewriting is therefore usually of dubious legality.) Since true software innovation is in the algorithms rather than the code they happen to be programmed with, the crucial innovations of computer science are only incidentally protected by copyright. To put the point bluntly, few software innovators would really care if someone copied their programs to read them, if they could prevent unauthorized execution of their programs. Yet controls on who may copy software effectively prevent, almost as an accidental side effect, unauthorized execution of the program. Copyrights are therefore effective protections for software.

Copyrights have been happily embraced by software owners. In many respects they are preferable to patents, even if patents were an option. Whereas patents are granted after a lengthy, expensive application procedure, copyrights are simply filed. Whereas patents extend for about seventeen years, copyrights last for about seventy-five years. There are, however, serious questions about the appropriateness of copyrights for the software industry. They may be too strong and may too effectively exclude new competitors from entering into the computer industry. This is the central issue in the debate between Duncan Davidson and Pamela Samuelson.

Davidson bases his support for software copyrights on an economic study of how the industry did react to the court decisions that established the availability of software copyrights. In Apple v. Franklin (714 F.2d 1240 [1983]), the courts upheld copyrights on programs controlling the internal functions of personal computers. (Franklin had blatantly copied these from Apple. Franklin challenged the validity of Apple's copyrights on several grounds, but the courts clearly favored Apple. Franklin abandoned the case after losing arguments in some preliminary injunctions, and the message went out to the software industry that copyrights were "in.") Following this decision, IBM sued several competitors for infringement of their copyrights on similar programs. Through a study of the pattern of these IBM suits, Davidson concludes that copyrights have had a positive effect on the software industry. He argues, in particular, that even though the copyrights are being used to protect methods for designing computer functions, copyrights are not so limiting as to block new entry into the personal computer industry.

Samuelson disputes Davidson's conclusion. She focuses attention on the tradition that copyrights should encourage public disclosure of the ideas contained in the copyrighted material, and thus encourage open exchange of scientific results and discourage trade secrecy. She argues that, to the contrary, software copyrights encourage secrecy and discourage open exchange. To understand her argument, we must recognize that software is usually published in unreadable, executable code. A researcher who wishes to understand the software must first put it in readable form. Sadly, the process of putting the code in readable form will usually involve copying the code and thus infringe the copyright. So we have a situation where attempts to understand the copyrighted material are infringements of the copyrights. This would not be so serious if the programs were publicly filed in readable form at the Copyright Office. But, in a special exception for the software industry, the Copyright Office permits software engineers to file their works with only small portions in readable form.

The debate between Samuelson and Davidson, then, is largely over the effect that software copyrights have on the public disclosure of new technology. Davidson argues that the industry has not been hampered, since qualified researchers can infer the nature of the software algorithms from the way programs work without needing to make unauthorized copies. Samuelson argues that software copyrights prevent researchers from studying each other's results, and are thus discouraging rather than promoting research into computer software.

Biotechnology and Patents

Among the products of modern biotechnology are organisms that produce valuable substances. To take a fanciful example, consider the possibility that a biologist, fiddling with the genetic makeup of a bacterium, creates an organism that eats ordinary trash and turns it into pure, consumable alcohol. The biologist may wish to exploit commercially the organism without worrying about the possibility that viable cultures fall into the hands of competitors. A patent on the organism would block commercial use of that organism by competitors, should they acquire it. Recent court decisions have opened up the possibility of such patents, contrary to a long tradition that denied patents on life-forms. As with copyrights on software, patents on organisms involve special accommodations for biotechnology within intellectual property law. (There had been a provision in patent law for "plant patents." But the organisms developed by biotechnologists are not plants in the sense of those patents, and the courts had taken the possibility of plant patents as an argument against rather than in favor of granting patents on organisms.)

In Diamond v. Chakrabarty, 447 U.S. 303 (1980), the Supreme Court recognized that some organisms may be patentable subject matter. If the organisms are naturally occurring, the researcher has made a discovery (discoveries are not patentable) rather than created an invention (inventions are patentable). But if the researcher has either (1) modified the organism through genetic engineering or (2) developed a pure strain of an organism that does not exist separated from other organisms in nature, then the researcher might have modified nature to a sufficient degree to have an invention in the sense required for patents. In both cases, the resulting organism is a nonobvious invention developed through the use of advanced biological and chemical techniques. However, there remain serious problems in the application of patent law to these new sorts of inventions. A. J. Lemin discusses problems dealing with the modification of organisms. Carl Cranor discusses problems arising out of the development of pure cultures.

Lemin focuses on the requirement that inventors disclose in their patent applications all technological details needed to produce and to make optimum use of the invention. The application must enable anyone skilled in the relevant technology to produce and use the invention without extensive experimentation or development. To appreciate this requirement, we must note that some inventions are commercially exploited outside public view. Although an invention that is a new part for an automobile cannot be exploited without making the part available on the market for competitors to study, a device for producing automobile parts may be exploited in the factory away from the eyes of competitors. If that device is patented rather than kept as a trade secret, its workings must be publicly disclosed. Organisms discovered by genetic engineering are, like production devices, often exploited out of public view. An organism, like our fanciful trash eater, may produce a substance that could also be produced in more traditional ways. If the substance shows no sign of the organism used in its production, it could be produced in secure factories without disclosing the existence of the new organism.

In their patent applications, genetic engineers must show how any skilled biotechnologist can develop the organism. The problem is that it is not easy to see how one would disclose the technology that permits quick and easy development of a genetically engineered organism. At present, a culture or sample of the essential genetic material of the organism is made publicly available. Even that resolution of the problem has problems. Lemin argues that to make cultures or samples of genetic material available is both too much (leading to loss of control over the organism) and too little (insufficient in itself to permit optimum use of the organism or genetic material). He shows that even though patent protections may have been extended by the recent decisions that cover modified organisms, the details of the accommodations of the patent law for this purpose continue to create obstacles to the use of these patents. These problems then leave the field of biotechnology largely without adequate protections.

Carl Cranor focuses on the patents on pure strains of organisms that do not exist naturally in that pure form. He explores the case of an apparently valuable beastie that was developed through investigations into matter found in the blood of a cancer patient. The patient's doctors were able to identify an organism in the blood, isolate it, and then develop a viable culture that can be maintained in laboratory conditions. That research established a culture of an organism that is not naturally occurring, and as such is a patentable strain of an organism. This organism produces valuable medical substances at a considerable rate and was sold to a commercial developer for several million dollars. A dispute then occurred between the patient whose blood contained the original source for the culture and the doctors who developed the viable strain. Each party claimed rights over the commercially valuable beastie.

Cranor views the dispute as a question of the duties owed by medical researchers to their patients and of the expectations of researchers to be able to profit from studies based on patient cases. The issues go far beyond the particular case to worries about the very nature of medical practice and research in an environment where new commercial opportunities have given rise to unexpected controversies over personal property. The fact is that the changes in intellectual property policy that accommodate genetic engineering have given rise to conflicts with prior notions of personal property. These conflicts remain a basic challenge to the intellectual property system.

7. Reverse Engineering Software under Copyright Law: The IBM PC BIOS

Duncan M. Davidson

This case study focuses on the impact of recent changes in the law affecting business and technology. The 1976 Copyright Act, which took effect in 1978, left open the question of whether all forms of software are copyrightable. A recent federal appellate court decision involving Apple Computer and the manufacturer of an Apple II "clone," Franklin Computer, held that all forms of software are copyrightable. IBM then sued three manufacturers and distributors of IBM Personal Computer clones. This case study analyzes the effect of this change in the law on IBM PC clones and its impact on developing technology. It concludes that the effect of the change is benign, that it is possible within copyright law to create clones, and that in an opportunity economy, business and technology can meet legal challenges such as this one with relative ease.

The spark that lit the computer industry was the concept of organizing the computer on logical, not engineering, principles (Fishman 1982). A computer designed logically can be redesigned with relative ease by programming, whereas one designed by hardwired engineering principles can be redesigned, if at all, only by painstaking rewiring. The breakthrough that allowed the development of a logical computer was the concept of programmable memory. The symbolic values of memory contain the logic of the computer. A change in these symbolic values changes the machine—a concept revolutionary to engineering.

These symbols comprise the programs and data (the *software*) of the computer. The marvel of software is that it is not only a set of symbols in memory, but the very driving force of a computer. Everything the computer does, and the order in which the computer does it, is determined by software. Software is part of the machine. Indeed, software and computer hardware are fundamentally interchangeable. All computer programs could be implemented either as digital logic circuits (hardware) or as volatile binary information in computer memory (software). Software is *both* engineering and symbolic writing in a computer language.

This dual nature of software causes untold problems with the law. Computer software is not well understood; is subject to many misconceptions; and it can be categorized over a whole range of possible subject matter from services to languages to new forms of hardware (Davidson 1983). Courts, lawyers, accountants, bureaucrats, regulators, legislators, and scholars continually try to fit software into existing legal and tax categories, but it will not fit. These categories take only one or the other side of the dual nature of software into account. Is software tangible (the engineering side) or intangible (the symbolic side)? Is software patentable or copyrightable? Is it depreciable as a capital asset or amortizable as an intangible asset? Is it goods under the Uniform Commercial Code or a service under general contract law? These questions extend endlessly into the many nooks and crannies of the law.

The concern of this case study is where copyrightable software ends and patentable hardware begins. Once a computer is understood as a logical machine, the computer designer can implement as software parts of the computer previously implemented as hardware. IBM in its Personal Computer put more into software than other comparable machines. This case study analyzes the legal problems that this engendered.

The Problem

Traditionally, it was permissible to reverse engineer manufactured products, unless aspects of those products were patented. Software is the first engineering product, however, which is copyrightable. It might not be possible to reverse engineer a computer program without violating the copyright. The question is whether copyright protection of software in some cases may upset the balance between patentable and unpatentable inventions by precluding this traditional right to reverse engineer unpatented manufactured products.

If so, copyright might give patentlike protection for software inventions that otherwise would fall short of the standards for patentability. In such a case, the safeguards in patent law would not have been followed: there would have been no examination to see if the software had sufficient novelty to deserve protection superior to other engineering products. Further, the protection period would last at least fifty years, longer than the seventeen-year protection period of patents. Fortunately, these fears about patentlike protection of software through copyright are not justified.

The IBM PC BIOS

This case study analyzes the question in one of its clearest circumstances. The most fundamental software in the IBM PC is the basic input/output system (BIOS). The BIOS contains a series of short routines that connect the various hardware elements into a logical structure. The hardware elements involved include the keyboard, the disk drives, the monitor or terminal, the printer, and various subsystems that interface with the central processing unit of the computer and these other hardware elements.

For example, when a computer program accepts input from a keyboard, the command or set of commands in the program that allow for input from the keyboard would be interpreted by the computer programming language to make a call to the BIOS. This call sets in motion a short series of computer instructions stored in the BIOS that route input in the form of keystrokes on the keyboard to the proper place in memory for further use by the original program.

Software for the PC—and in particular, operating-systems software and computer languages—use these routines stored in the BIOS to interface with the logical computer. The BIOS allows the real computer to act like the logical one. This level of abstraction enables a company such as IBM to change certain hardware elements of a computer, and to make corresponding changes in the BIOS, without having to change any other software that relates only to the logical machine and not to the actual machine.

Although the BIOS is becoming a standard name for this part of the IBM PC, it represents a more generic way of designing a computer. A computer need not have a BIOS, and it could also have a BIOS that is either shorter or longer than the IBM PC BIOS. It is within the computer designer's options to create computer architecture that has more or less of its fundamental components in software as opposed to hardware—that is, in the form of a BIOS as opposed to actual hardwiring. What the computer gains by a BIOS or its equivalent in flexibility it often loses in efficiency.

Compatibility and Clones

In order to reverse engineer a PC and make a competitive machine, it is necessary to make the competitive machine compatible with the original machine. *Compatible* in this context means that the copy of the computer (the clone) must have the same logical structure as the original so that all programs that run on the original machine will also run on the clone. If a PC clone had all of the same hardware elements as the IBM PC, the BIOS could be identical, and that would create a compatible machine. If the hardware elements of the clone were different in certain respects, the BIOS would be written differently in those respects. The key to compatibility is understanding the logical machine—understanding how the BIOS works with the hardware. The more identity there is between the original PC and the clone, the more identity there will be of necessity in the BIOS.

The Issue Narrowly Stated

Assuming that copyright law protects the original IBM PC BIOS, it is not permissible to make an exact clone using the exact BIOS. This would violate copyright law just as making an exact copy of a book would violate the copyright of the book's author. Instead, the BIOS would have to be rewritten in a fashion that would not violate the copyright of IBM and yet would still create the same logical machine, the same compatibility.

Rewriting a BIOS may not be easy. Copyright protects more than the exact way something is written; copyright protects expression, not embodiment. Copyright protects the writing, not the book; similarly, it protects the program, not the exact code in which it is implemented. Just as expression in a book does not consist of the precise language in a book, but of the specific interrelation of plot, character, theme, literary techniques, and other literary elements of the story, the expression in software does not consist of the exact written code, but the specific interrelation of structure, routines, algorithms, programming techniques, and other program elements (Davidson 1983, pp. 659–660).

If it is not possible for a PC clone manufacturer to write a different BIOS and achieve the same compatibility, then the traditional balance in patent law between protectable and unprotectable engineering may be upset. To put it in another way, if a company such as IBM had designed a personal computer without a BIOS or other copyrightable type of logical structure, then there would be little question that the clone manufacturers could copy the PC. By simply putting some of the architecture of the PC into software, and using copyright law, such a company may have found a technique to give patentlike protection to its computer architecture when no patent was actually issued. As this case study indicates, however, that hypothetical company has not found such a technique. As a factual matter, PC clones have been created that apparently have BIOSs that do not infringe IBM's copyright.

Legal Background and History

The question of whether copyright protection of fundamental operating-system programs in a computer can give patentlike protection to a computer's architecture has arisen several times. The software copyright cases are best understood when the technology of programming is appreciated.

Source code is the name of software when it is first written. Today, most source code is written in higher-level languages—so called third-generation languages—such as FORTRAN, BASIC, C, and COBOL. These languages have many of the attributes of English that make writing programs in them much easier than using second- or first-generation languages (Davidson 1983, pp. 619–624).[1] There has been little dispute that source code is copyrightable, even when written in a first-generation language. In this case, the programmer has written in a human language that consists of symbolic meaning, as is the case with other literary works.

Questions of copyrightability arise when the source code is translated into machine code through the use of programs called assemblers, compilers, or interpreters (Davidson 1983, pp. 699–701). The resulting machine code is often called *object code*. When combined with other programs, it can create a load module, which is an actual machine-readable program that the computer executes. The problem with object code is that it is not meant to be read by people (although it theoretically could be), but to do work. Arguably, this is so far beyond the pale of copyright that it falls within the realm of patentability. This issue has been scrutinized from different angles depending upon the particular facts of the cases involved.[2]

Software Copyright Cases

Defendants against copyright infringement in these cases have argued that while source code is copyrightable, object code is not since it is unintelligible and used merely to operate a computer, not to be read by people.[3] Section 102(a) of the 1976 Copyright Act, however, allows copyrightable works to be embodied in unintelligible forms as long as the work can be made intelligible "with the aid of a machine or device."

Defendants have further argued that a program put in the form of a read-only memory (ROM) chip was not copyrightable because it was a piece of hardware.[4] ROMs look, feel, and act like other hardware elements. Indeed, fundamentally all digital logic circuits in a computer are simply a series of gates connected together in a specific way to perform logical operations. These gates can also act as memory locations. A ROM is a form of this type of memory. One could argue that a ROM's sole function—even more than object code's—is to perform work and not to be an embodiment of copyrightable expression that somebody would eventually read. But the 1976 Copyright Act makes it clear that the medium is not the message, that the form in which copyrighted work is embodied is not dispositive.[5] A ROM could just as well hold a novel as a computer program.

The final argument of such defendants has been that operating-system software was different from other software because it was so closely entwined with the architecture of computer hardware.[6] Programs are generally divided into three categories: applications, operating systems, and utilities. Applications accomplish something, such as word processing or accounting. Operating systems are a type of BIOS at a higher level; they cause the computer to operate properly and provide an environment in which an application can run. Utilities are programs that are used largely for the creation of new applications and to make certain repetitive computer operations easier. Operating systems are written in the same way that applications are, and the line between applications and operating systems is often hard to draw. It would be possible to put a number of the aspects of an operating system into an application directly, or vice versa, depending upon the type of computer and the style of the programmer.

All of these arguments have failed in the more recent cases. The defendants were fighting a losing battle. The question is not that adding object code to the copyright system is somehow grafting inappropriate subject matter into the system; instead, the proper argument is the opposite. It is clear that source code is copyrightable, and it is clear under the 1976 Copyright Act that machine-readable source code is copyrightable. On what basis should object code be removed from the copyright system? Congress evidenced a clear intent in 1980 to include it when it amended the 1976 act explicitly to include computer programs.

Based on current case law, it is clear that all forms of software in all embodiments are copyrightable. It does not matter whether they are source code or object code, embodied in floppy disks or ROMS, accomplish operating system functions or applications, or any other similar distinctions. Nevertheless, attempts to narrow the definition of copyrightable subject matter on the basis of the form of the embodiment reflect a more fundamental problem with copyright of software than has been argued in these cases.

The Policy Question

Underlying these cases is the dual nature of software: it is engineering, but it is expressed in a language, in a symbolic fashion. It is the first language that by itself can operate a machine. It is also the first copyrightable work that is not expected to be read. Other copyrightable works are designed to be used by people, even if the work itself may be expressed in some unintelligible way. Record albums are unintelligible, but when they are used, the music (which is copyrightable) is played. Videotapes have illegible magnetic pulses on them, and yet when they are used, the copyrightable work is displayed. Books, of course, are designed to be read. Even encrypted literary works are copyrightable, because there is the intention that they will be decrypted and read by the proposed user of the work. In contrast, when software is used on the vast majority of computers, the program itself—the copyrightable work—is not intended to be read or otherwise enjoyed by the computer program user. What the user reads and enjoys is expressed in the *user interface*.

The question underlying the software copyright cases is whether copyright law should protect those forms of software which are intended only to be operated inside a computer, not actually read by a person. This question must be answered as a policy choice, not in the abstract. Removing that type of software from copyright protection would leave the vast majority of software being marketed in this country unprotected and vulnerable to piracy. For economic purposes, the nation would have to create a new type of intellectual property system to protect substantially the same interests in software that are now being protected by copyright law.

A number of so-called sui generis systems of protection have been proposed (Davidson 1983, pp. 760–784). With the advent of the new Semiconductor Chip Protection Act, other software protection systems similar to that created by the act may be proposed.[7] In general, these proposals have been based on copyright principles. The major differences from the current copyright system are: (1) a shorter period of protection, (2) a change in formalities (usually more disclosure), and (3) a drawing of the line between protected and unprotected aspects of software in different places. Some proposals would be more protective, encompassing the design of the software as well as its expression, while others would be less protective, making it easier for competitors to reverse engineer certain types of software and create competing versions. None of these proposals avoids a fundamental problem: drawing a clear line between protectable and unprotectable aspects of software.

Drawing that line is a policy question. In the context of this case study, this question can be faced in a simple, dramatic fashion: Does protecting software in the nature of a BIOS give too much protection to computer system manufacturers, upsetting the traditional balance between patented and unpatented works?

Apple v. Franklin

This policy question was most clearly addressed in the case of Apple Computer v. Franklin Computer Corp.[8] Franklin Computer made an Apple II clone. It copied the basic operating-system programs of the Apple II, including those programs which perform some of the same functions as the BIOS. When Apple sued, Franklin defended by using all the available arguments that had been used previously, plus two others. The previous arguments that were repeated were, first, that object code was not copyrightable; second, that the programs that were embodied in ROM were not copyrightable; and third, that the programs that constituted the basic operating system of the Apple computer were not copyrightable. The two new arguments were first, the "communication" argument—that machine-readable programs are not traditional copyrightable works because when they are used they do not communicate the copyrighted expression to people; and second, the underlying policy question—that copyright protection of these operating-system programs would make it impossible for Franklin to create a compatible clone.

As had happened in other cases before it, the court on appeal discarded distinctions between types of programs, based upon whether the software was object or source code, recorded on magnetic media or on ROM, and so forth. Nor did it view the communication argument as dispositive. Instead, the court faced up to the policy question directly. It noted that copyright law protects only expression and not underlying ideas. It characterized Franklin's defense as saying that copyright protection of certain types of programs protects not only the particular way the program is written (expression) but also the underlying ideas (the manner in which a BIOS, for example, creates a logical computer).

The court answered this policy question with a factual analysis: as long as these fundamental programs could be written in several ways, and the other ways would not violate the copyright of the original way, then copyright protection is not too strong—it does not cover both the ideas as well as the expression. On the other hand, if there is only one way, or a very small number of ways, to write the program, then the expression merges with the ideas, copyright will not be enforced, and a clone maker will not be precluded from making the copies.[9] In either case, copyright does not protect ideas, only expression. The appellate court therefore remanded the case to a lower court for analysis of the factual question. The case was later settled before the analysis was accomplished.

IBM Sues Clone Manufacturers

After this case, and coincidentally after IBM had been released from the pressure of an antitrust lawsuit, IBM turned its massive legal talent into a number of areas to protect its copyright interests. Among other actions, it sued three of the PC clone manufacturers and distributors, alleging that they copied its BIOS.[10] All three—Eagle Computer, Corona Data Systems, and Handwell Corporation—settled. More recently, IBM successfully enjoined distribution of PC clones in Canada.[11]

The three American companies producing clones suffered a certain amount of disruption. As a result of the settlements, they apparently would have to rewrite their BIOSs to make them noninfringing of IBM's BIOS. Although it is difficult to ascertain by telephone conversations with the attorneys of these companies the extent to which the lawsuit caused problems, from certain documents that have had to be filed for federal securities purposes, the effect of the lawsuits can be seen as dramatic. Consider the following disclosure filed by Eagle Computer:

> Third quarter net sales were affected by a suit brought against the Company by IBM which alleged a copyright infringement by Eagle of its BIOS—a program inside the computer which controls the computer's inner workings. Although there was no payment made as a result of the suit and the Company admitted no guilt, as part of the settlement Eagle

> agreed to cease shipments during March while a new BIOS was completed and tested. As a result, shipments were considerably attenuated during what is usually the heaviest shipping month of a quarter, the third month. Additionally, shipments of a new product, the Turbo PC were delayed due to engineering resources being redeployed in order to finish the BIOS.
>
> As a consequence, net sales for the third quarter were $10,003,000, nearly half the previous quarter. Due to the lower sales and other factors described below, the Company experienced a loss of $9,865,000 for the third quarter.
>
> In the highly volatile personal computer market, the announcement of the loss by the Company caused many prospective customers to select other vendors with similar products who had not reported such a loss. As a consequence of this and a general softening of the personal computer market beginning in late spring of 1984, sales in the fourth quarter showed a significant decrease from third quarter sales. Net sales were also affected by product returns of $3,200,000 primarily from customers who refused to pay for products previously ordered and received. Net sales in the fourth quarter amounted to $4,961,000. The loss for the quarter was $17,090,900. (Securities and Exchange Commission, Form 8-K, 29 October 1984, p. 4)

Franklin's Problems

Franklin Computer also had problems after settlement of its lawsuit with Apple Computer. Apparently, it had been rewriting many of the Apple operating-system programs while the lawsuit was pending; shortly after the lawsuit concluded, it was able to come out with a second version of its computer. Unfortunately, its rewritten programs were not 100 percent compatible, and it had serious problems in the marketplace because many programs written for Apple II computers would not run completely on the Franklin computer.

This problem was exacerbated because so many Apple third-party programmers had used the insides of the Apple computer to make their programs more efficient. In effect, they had not used the logical computer presented to them by the basic operating-system programs, but had gone a level deeper and used alternate methods for manipulating the actual hardware. Franklin not only had to create the same logical computer, but also had to provide for these alternate channels for using the underlying hardware elements that the more popular programs were using. (This problem is much less prevalent in the IBM PC world, because about one-third of the market has bought PC clones, and third-party programmers take great steps to write programs that will work on the vast majority of clones as well as on the original IBM PC).

Franklin was later able to rewrite its operating-system programs to make them more compatible with Apple programs, but financial difficulties forced it into a reorganization proceeding under bankruptcy law, and its long-term viability is uncertain.

Robust Commercial Response

There is a whole other side to the story. Business and technology appear to be much more deft at moving around legal obstacles than might otherwise be thought. First, many other PC clone manufacturers were not sued. One can infer that they were able to write their BIOSs in a noninfringing manner. These companies include Compaq, Tandy (Radio Shack), Olivetti (AT&T), Zenith, and Texas Instruments. Second, at least two of the three PC clone companies that were sued have been able to produce a BIOS that apparently has satisfied IBM as noninfringing. Third, and most interesting, independent third-party BIOS companies have arisen, the most popular of which is Phoenix Software Associates in the Boston area.

The economic effect of the IBM lawsuits is now becoming clear. Those companies which did not produce their own BIOS but simply copied the IBM BIOS have been forced to start over and rewrite their BIOS. This has increased the cost of producing a compatible IBM PC clone for some of the competitors. For those companies, simply copying the BIOS would have been a windfall. If the IBM PC had been largely hardwired without a BIOS, they would have had to spend additional time reverse engineering to determine how that hardware worked rather than simply copying a BIOS. Now that they and the other companies producing clones have been forced to rewrite their BIOSs, their development costs may now exceed those which they would have had in simply reverse engineering a hardwired IBM PC.

Indeed, this cost increase can be somewhat traumatic. When IBM first created its PC, it could write its BIOS for a relatively small amount of money. It takes much more money to create a rewritten BIOS in the proper fashion, since much testing and experimentation has to be done to determine that compatibility has been achieved. Nevertheless, the barriers to creating clones have not been that great. A large number have been created, and some of them have achieved significant returns on their investment (the most dramatic examples being Compaq and Zenith).

Further, the third-party BIOS companies have lowered the cost of creating a BIOS even more. These companies spread the development costs for creating a clone over a large number of customers, thereby lowering the cost to each individual. The ironic result may be that although IBM has significantly increased the barriers to entry of competitors who might otherwise have tried direct copying, a free-market economy has responded, through the advent of the third-party BIOS companies, ultimately to lower development costs to a reasonable level—and to make it easier to enter the clone game. It is impressive how fast the free-market system has acted in this circumstance to increase competition, diversity, and opportunity.

Discussion and Analysis

These facts lead to a conclusion that copyright protection is not too strong and does not give patentlike protection to computer architecture. The discussion is still in the middle of these circumstances, however, and certain qualifications are in order.

The Problem of Hooks

The Franklin situation illustrates one qualification. Franklin appears to have had a much harder time creating a compatible system than the companies making IBM PC clones. The reason may be due largely to the third-party software companies that went below the logical design of the Apple computer and used or created other channels (*hooks*) for using hardware elements. Indeed, the hobbyist mentality that led to the creation of many of the original Apple II programs is one in which the most clever programs are the best, not the ones that are written by following the rules of the logical computer.

The policy concern is whether this circumstance will ever repeat itself. A very different set of circumstances occurred with the advent of the IBM PC. Cloning was then seen as a likely occurrence, and many of the programs for the PC were created after clones had first appeared. As a consequence, for economic reasons the third-party programmers wanted to avoid as much as possible going below the logical machine and hindering their ability to be useful on the broadest spectrum of computers possible.

Closed vs. Open Architectures

The second qualification is illustrated by IBM itself. IBM's entry into retail marketing with the PC was a high-risk gamble. It has succeeded beyond most people's wildest speculation. IBM borrowed many of the aspects of the Apple II computer to enhance its likelihood of competing with the PC, including an open architecture that encouraged PC clone manufacturers, third-party add-on board manufacturers, and third-party programmers to develop products around the PC. Now that it has established a very large market, and has once again flexed its marketing muscle, there is no assurance that it might not return to a closed architecture in the future releases of the PC. Among other problems, a closed architecture may encourage third-party programmers to write very specifically for the new PC, and this may make it extremely difficult for PC clone manufacturers to create complete compatibility.

IBM's intentions can be gauged by the design of its second-generation of PCs, the "PS/2" line. IBM retained the open architecture but made it more difficult to clone the PS/2 by using the full range of intellectual property protection (patents, copyrights, trade secrets, and trademarks). Nevertheless, the PS/2 has been cloned. In addition, clones of the original PC line are still selling vigorously, and market forces may limit IBM's ability to impose the PS/2 standard, let alone a fully closed architecture standard.

IBM may also be inhibited from converting to a fully closed architecture because of the threat of antitrust litigation. In recent decisions, Data General, for example, was found to have engaged in practices that were illegal under the antitrust laws when it attempted to restrict the ability of persons making clones of its Nova computer.[12]

Validity of Inferences

A third qualification is whether too great an inference is being drawn from the fact that IBM has not sued other clone manufacturers. First, IBM may have been reluctant to go too far in suing clone manufacturers because of the possibility of antitrust actions against it. Second, IBM may have taken a very conservative view as to the scope of copyright. It may have sued only those clone manufacturers who made duplications of the IBM BIOS. Recent copyright cases may extend the scope of copyright protection much farther than IBM believed several years ago, and IBM may in the future begin taking action against other PC clone manufacturers.

Conclusion

Speculation about IBM's intentions is fraught with difficulties and is not a sound basis for structuring a legal protection system for software. These qualifications do not provide a strong enough basis for challenging the conclusions of this case study.

Further, despite these qualifications, a fundamental factual and legal circumstance remains. If it is possible to write a BIOS in several ways, then by doing so a clone manufacturer can avoid a lawsuit. If it is not possible to write a BIOS in more than a few ways, then longstanding policies of the copyright laws involving the difference between protecting expression and protecting ideas can come into play and again protect the clone manufacturers. If a dominant company illegitimately restricts cloning, then antitrust laws apply to protect the clones.

Recommendation

The key to creating a compatible clone is to follow safe procedures in writing the BIOS. The currently recommended procedure is to use a *clean room*. In general, a program will violate another's copyright if it is "substantially similar" to the original program. If it can be demonstrated that the programmers of the new program never had access to the old program, however, even if they create a program that has similarities, then they would not have copied. The clean room is a technique to avoid *access* despite creating *substantial similarity*.

In a clean room, a group of persons who investigate how the target BIOS works is separated from the group of programmers who actually write the new BIOS. In this manner, the programmers never have access to the original BIOS.[13]

SAS v. S&H

Clean rooms must be used with care. The programmers in the clean room must be fed only the design information of the original program (the ideas), not any specific elements of programming itself (the expression). In the recent case of SAS Institute v. S&H Computer Systems,[14] S&H attempted to copy certain programs of SAS using a clean room, but it was not clean enough. The original software was a series of scientific routines that were quite popular with users of IBM mainframe computer systems. The manufacturer of the clone wished to recreate similar routines for use on Digital Equipment Corporation VAX computers. S&H acquired the source code of the IBM mainframe version of the software, and purportedly had a design staff carefully separate the ideas from the source code and submit them to a clean room of programmers. Nevertheless, the clean room failed, and S&H was found liable for a copyright violation.

The reasons for this failure are subject to interpretation. For example, the mere fact that S&H acquired the source code of SAS with the express purpose of making a copy may have been strong evidence against them; they appear to have acted with larcenous intent. However, the key finding of the court was that certain nonfunctional, vestigial elements of the SAS programs were found to be replicated in the S&H programs. These apparently were elements that were functional in older versions of the SAS software, but since that software had been revised, they had become nonfunctional. As with many other large software products, no one thought it worthwhile to clean up the old source code so long as the new programs worked. Apparently, the persons analyzing the SAS software submitted to the clean room more than merely the logical design of that software; they also submitted specific elements of programming code (the expression of the original software). A further factual circumstance held against the copier was that the time period for creating the copy was suspiciously short.

Copyright law in these circumstances is evolving into a misappropriation law, which would protect the head start granted the first to produce a certain type of software, because of the dedication of time and expense required. Perhaps if someone were to reverse engineer software in the proper way, there would be a minimum amount of time that this would take. Too hasty copying of someone else's software will lead to an inference that more was taken than merely the ideas and design. On the other hand, an injunction may only issue for the period of time it would take for the guilty party to rewrite the software in the proper way (Davidson 1983, pp. 704–705, 720–721). Thus, this evolution to a misappropriation type of law does not create patentlike protection. While it raises the costs of someone's making a competing version of software, it also in effect forces that party to innovate, to do the new version in a better fashion than the old version.

The clean room is one way to accomplish the innovation even while creating compatible software. Nevertheless, the clean room staff should refrain from rushing ahead to create the competing version and appropriating too much of the expression of the original software. A clean room staff must be very careful that it is receiving only ideas and not expression of software, logical design and not specific algorithms.

Decompilation

A second caveat regarding the use of the clean room is how one gains access to the design of the target software. The PC BIOS was published, but most source code is not. The SAS software was acquired through an appropriate license under which the source code was marketed, but many programs are marketed only in object-code form.

The common method for overcoming these obstacles is to decompile or disassemble the object code. Source code is turned into object code by a program that compiles or assembles it from the higher-level version to the machine's version; a program can also reverse this process, although not completely, by disassembling or decompiling. The new copy of the software created by disassembling or decompiling, however, is an unauthorized copy that may violate copyright principles. It may be that if one were to take the fruits of the illegal decompilation and create competing software, even through a clean room, the court may find that the damages caused by marketing the competing software relate back to the original act of decompilation. In other words, if the clean room was tainted by the unauthorized decompilation, then all resulting products might be subject to copyright injunction and damages.

This is an area of law that is unsettled and at the forefront of the next round of software copyright issues. In effect, it restates the original question, moving from a conflict between patent and copyright law to a conflict between trade secret and copyright law. Copyright is supposed to protect expression and not underlying ideas. Software is the only copyrightable work that is normally marketed in a form in which the ideas are hidden from the casual user. If a person is not allowed the right to decompile or figure out how software works (that is, to determine what the underlying ideas are), in effect trade secret–like protection has been given to subject matter that is normally not a trade secret. Whether this is appropriate remains to be seen.

In the meantime, the safe course is not to decompile software but to attempt to figure out its logical design through other means. For example, if the software is in a ROM, a series of signals can be input to the ROM, and the output measured. Through this process, it is possible to determine how the ROM works without actually looking at the software that is embedded in the ROM. Similarly, object-code software could be put through all the various transactions that it can accomplish, and the manner in which it accomplishes the transactions can be analyzed. In this fashion, many of the underlying algorithms and procedures of the software can be ascertained. In both cases, the logical design of the underlying software (the ideas) can be determined. The process is not always easy, and sometimes it takes more time than it would take simply to attempt to write competing software from scratch. If the desire is to achieve the same functions as competing software, this process is not necessary. If the desire is to achieve complete compatibility with existing software, then this process, although lengthy and arduous, will eventually lead to that goal.

Policy Analysis

There are a number of different interests in the area of software protection. They can be categorized in three general areas. Most protective would be software vendors, for obvious and nonobvious reasons. The least protective would be copyright traditionalists, persons who believe that software is not appropriately grafted onto the copyright system because software is not meant to be read but rather to do work in a computer. Skeptical of both of these interest groups would be the contrarians, persons who wish to promote diversity, opportunity, and innovation rather than protect the entrenched positions of major software vendors or protect the pristine beauty of traditional copyright and patent systems. The interests of each of these groups will be discussed in turn in the context of this case study.

Protectionist

The *protectionists*—the software vendors—originally marketed their software under trade secret licenses. Until 1978 machine-readable forms of software were not protectable by federal copyright laws, and trade secret law was the only feasible means for mass-marketing software. In this effort, the protectionists were stretching the traditional use of trade secret law, which normally was used (1) to protect secrets against disloyal employees and (2) to allow the limited marketing of know-how to certain select customers. In contrast, software was marketed to a nonselect group (as long as they would sign licenses) without, in many cases, disclosing know-how embodied in the software.

After the 1976 Copyright Act took effect in 1978, and again more clearly after amendments to it in 1980, there was much less need to use trade secret licenses in marketing software. The protectionists continued to do so, however, for two reasons. First, there were many perceived inadequacies in copyright law, the most important one remaining being the First Sale Doctrine embodied in Section 109 of the 1976 Copyright Act. Under this doctrine, once a copyrighted work is sold, its further use cannot be restricted. The buyer could resell it, destroy it, rent it, or revend it in other fashions. The concern of software vendors was that the renting of software would allow the pirating of software. The renters would simply make a copy of the program, photocopy the documentation, return the original at the end of the rental period, and continue to use the program without having to pay the software vendor any royalties.

The second reason for continued trade secret licenses was to keep traditional trade secret protection in force against employees and those select customers to whom know-how and ideas in the software were actually revealed. The fear was that if software were mass-marketed, and especially if such actions as decompilation were permissible, then the trade secrets in the software would have been disclosed to the public. In that context, even traditional trade secret law would allow employees and select customers to use those disclosed trade secrets with impunity.

In the context of the PC BIOS, neither of these interests of software vendors seems compelling. The BIOS is best implemented as a ROM, and it would make no sense to rent it and pirate it. The trade secrets in the BIOS are minimal, and the threat to disloyal employees is somewhat irrelevant. When a computer system like the IBM PC achieves success, the attractiveness of copying the BIOS is not in learning trade secrets but in achieving the same compatibility.

Traditionalist

The *traditionalists* embody three related interest groups. One group does not like to see copyright corrupted by something like software, in which the copyright expression is essentially kept secret when it is marketed. Another group does not like to see the patent area infringed upon by engineering products like BIOSs, which are simply new ways of accomplishing what used to be done by hardwiring. A third group would not like to see trade secret law overly extended, but does want the ideas in software disclosed to the public.

All three groups would agree upon the following positions. The fundamental bargain made for either patent or copyright protection is disclosure to the public in return for a monopoly of either limited duration (for patents) or limited scope (for copyrights). Software should be protected against copying types of violations, but this can be best accomplished in the context of a new system of protection, a sui generis system that is neither copyright nor patent but something with the best of both.

A hypothetical sui generis system that comprises all three of these interests would be as follows. There would be a registration system that would include a deposit of the complete source code and any related information that helps people understand it. This would be immediately revealed to the public. There would be a patentlike protection system with a very short protection period (three to five years) for the system design, the ideas, and the algorithms being implemented by the software. There would be a copyrightlike protection system for a longer period of time (twenty to twenty-five years), protecting the exact or virtually exact embodiment of the program itself, but not reverse-engineered derivative works.

Contrarian

The *contrarian* seeks to optimize the software industry and is not concerned with formalistic considerations of how software copyright may corrupt the copyright system or infringe upon patent territory. The contrarian is concerned solely with empirical and functional considerations—what works.

The contrarian would use this case study to refute the major argument of the traditionalists: that copyright of software gives software patentlike protection. If this were so, then no IBM PC clones should exist for the seventy-five years of copyright protection of the original IBM PC BIOS. Their argument reminds the contrarian of the paradoxes of Zeno. Zeno demonstrated by logic that motion was impossible. Essentially, he stated that before one could go any set distance, one had to go one-half of that distance; and before one could go that one-half distance, one had to go one-half of the half distance; and so forth. The result is an infinite regression, and the person could never take the first step. Zeno's logic was rebutted empirically: When Diogenes, a famous Cynic, first heard of these paradoxes, he is reported to have stood up and walked away. Similarly, the traditionalists' logic is rebutted by PC clones. There is simply no basis for believing that copyright of software, even of software fundamentally entwined with the hardware of a computer, sets up such large entry barriers that patentlike protection results.

The contrarian is skeptical of the traditionalist concern that this vast body of copyrighted expression—source code—under the current regime will never be revealed to the public. Who will benefit from such revelation? The software industry—which would seem to be the best beneficiary of promoting the art of software through reading of the actual commercial literature—does not want this. The main cry for this appears to come from academicians. Is the academic pursuit of computer science at all enhanced by the vast body of mundane commercial software that has been produced over the last twenty years? There is no dearth of literature among computer scientists about their own ideas for advancing the art of programming. Would they really run down to the Copyright Office to read the reams and reams of source code that would be registered under the hypothetical sui generis systems? The most likely parties to do so would not be academicians but competitors wishing to appropriate ideas and quickly create competing versions.

Indeed, the very bargain the traditionalists assert as the basis for copyright—disclosure for monopolistic federal protection—may no longer be the sole bargain under the 1976 act, even though there is legislative history and there are cases under the 1909 Copyright Act that support it. The 1976 Copyright Act expressly extends protection to unpublished works. To that extent at least, the bargain is not struck to gain disclosure in exchange for protection, but to create a consistent federal system for all copyrightable authorship. There is an excellent argument that mass-marketed software is not published in the traditional sense precisely because the expression is not being revealed to the public. Is publication the act of mass-marketing or the act of disclosing the literary work to the public? These issues have never been considered before, because they never had to be. It is the unique nature of software to be that first copyrighted work which is mass-marketed without its ideas being published. The contrarian would instead argue that the *new* bargain being made by the 1976 Copyright Act is between *creation* of copyrighted works and federal protection.

The contrarian would also be skeptical of the protectionist points of view. The software industry began in a state of confusion over protection. The lack of clear standards for protection has not inhibited its rapid growth. As an empirical matter, it is still unclear what is the best way to optimize the software industry. Perhaps enforcing the protectionists' views would do more to entrench existing interests than to promote innovation and creation of new software companies. The rental concern may seem a logical argument, but it is refuted by empirical evidence. Is there a software rental industry? Until one develops, it is perhaps best not to take action to prevent it.

The trade secret concerns of the software vendors may also be more hypothetical than real. Could an employee steal trade secrets without also violating copyright? If there are such fundamental secrets, are they indeed jeopardized by mass-marketed software?

Just as mass-marketed software arguably is not "published" for copyright purposes, it may not be "disclosed" for trade secret purposes. In general, when a product is purchased in the free market, the buyer can tear it to pieces with impunity to figure out how it works.[15] One is therefore free to appropriate any trade secrets that may be discovered by reverse engineering (unless, of course, the use of such information infringes upon a patent or other right).[16] If the secrets are easy to ascertain from the product, the mere sale of the product is deemed to be disclosure of trade secrets.[17] If the secret is not so obvious, however, the buyer must actually go through the process of reverse engineering before the trade secrets are revealed. More precisely, injunctions against appropriation of trade secrets will extend for at least the amount of time it would have taken if such a defendant had gone through that process of reverse engineering.

When applying these principles to mass-marketed, machine-readable software, a synergy between copyright and trade secret law takes place. The very act of mass-marketing machine-readable versions of software does not in itself reveal any trade secrets. They are not visible from mere inspection of the product, but must be ascertained by a difficult process of reverse engineering. To accomplish this reverse engineering of software, the software must first be copied into human-readable form, and if it is in object-code form, be decompiled or disassembled to a more human-understandable form. One can argue that the making of any printout or other human-readable form of the software, including decompilation of object code, creates a new "copy" or "derivative work," which is unauthorized under the 1976 Copyright Act.[18] In some contexts, decompilation may be "fair use" allowable under the act. It is unlikely, however, that the fair-use doctrine would allow decompilation for the purpose of either publicly disclosing the software or creating competing versions of the software.

Consequently, the contrarian would argue that the current copyright system is very close to the theoretical optimal system for promoting the software industry. The IBM PC situation is a stirring example of this. There was no lack of competitive PC clones: in many cases they were less expensive, faster, and fuller-featured than the original. Copyright was strong enough to require the PC clones to go through the process of recreating the BIOS, and yet was not so strong that it created patentlike protection. Further, theoretically the trade secret interests of software vendors could be respected. This would be possible even when mass-marketing machine-readable software only under copyright protection because decompilation for competitive reasons would not be permissible.

In effect, under the current system software should be viewed as a black box. The insides of the black box remain relatively secret, but completely available for reverse engineering are the input, output, user interface (commands, macros, icons, and "feel"), and data-set format. Any person wishing to determine how the software works can largely find it out simply by watching how data that is put in is transformed into the output. Companies can create competing versions of the software—even compatible versions—by creating their own black box to have the same input, output, interface, and data format. These elements are not copyrightable unless they rise to the level of a video game in audiovisual complexity.[19]

Thus, the current system does not remove reverse engineering of software, but simply changes the way it is accomplished. Until it can be shown that this type of reverse engineering is not economically feasible, there is no reason to make any further changes in software protection. The performance of PC clone companies demonstrates this conclusion.[20]

Notes

1. First-generation languages are the direct machine code, sometimes programmed in series of 1's and 0's and sometimes in octal or hexadecimal form. Octal takes every group of three binary 1's or 0's and makes them into a number between 0 and 7, and hexadecimal takes every group of four binary numbers and makes them into a number between 0 and 15 with decimal numbers 10–15 represented by A–F.

Second-generation languages are assembly languages, one step removed from machine languages. In this case, the binary, octal, or hexadecimal code instructions to a computer would be represented in mnemonics. For example, if "1101" represented the operation of addition, then the mnemonic in assembly language may be "ADD." A program called an assembler would translate each ADD (or other mnemonic) into 1101 (or the other binary code of the machine language).

2. The common law, and the interpretation of statutes, proceeds on a case-by-case basis, not on an abstract legal basis that considers an issue in its entirety. It is difficult to foresee all of the ramifications of a broad decision, but it is much easier to see the effect of a narrow decision based on a narrow set of circumstances. As a consequence, the common law develops more slowly but with more accuracy than a broad-brush approach.

3. GCA Corp. v. Chance, 1982 Copr.L.Dec., Par. 25,464 (N.D. Cal. 1982); Apple Computer v. Formula Int'l., 562 F.Supp. 775 (C.D. Cal. 1983), *aff'd*, 752 F.2d 521 (9th Cir. 1984); Williams Elec. v. Artic Int'l., 685 F.2d 870 (3d Cir. 1982).

4. Stern Elec. v. Kaufman, 669 F.2d 852, 856 (2d Cir. 1982), *aff'd*, 523 F.Supp. 635 (S.D.N.Y. 1981); Midway Mfg. Co. v. Artic Int'l., 547 F.Supp. 999, 1012—1013, (N.D. Ill. 1982); Tandy Corp. v. Personal Micro Computers, 524 F.Supp. 171 (N.D. Cal. 1981); Data Cash Systems v. JS&A Group, 480 F.Supp. 1063 (N.D. Ill. 1979), *aff'd because of lack of notice*, 628 F.2d 1038 (7th Cir. 1980).

5. Apple Computer v. Franklin Computer Corp., 714 F.2d 1240 (3d Cir. 1983), *rev'd*, 545 F.Supp. 812 (E.D. Penn. 1982).

6. Idem; Apple Computer v. Formula Int'l., 562 F.Supp. 775 (C.D. Cal. 1983), *aff'd*, 752 F.2d 521 (9th Cir. 1984); Hubco Data Products Corp. v. Management Assistance, Inc., 219 U.S.P.Q. 450 (D.C. Ida. 1983) (hereafter cited as Hubco v. Management [1983]).

7. Public Law No. 98-620, 90th Cong., 2d sess., 8 November 1984, 98 Stat. 2347, Title III, Semiconductor Chip Protection Act of 1984.

8. 545 F.Supp. 812 (E.D. Penn. 1982), *rev'd*, 714 F.2d 1240 (3d Cir. 1983).

9. This principle is established in Morrissey v. Procter & Gamble Co., 379 F.2d 675 (1st Cir. 1967). See Continental Gas. Co. v. Beardsley, 253 F.2d 702, 705–706 (2d Cir. 1958) (noting that the copyright is valid even if the expression largely captures ideas, but that liability due to infringement beyond exact copying is difficult to establish).

10. IBM Corp. v. Handwell Corp., No. C84-20039 (N.D. Cal., filed 20 January 1984) (importer of PCs from Taiwan); IBM Corp. v. Corona Data Systems, Civil No. 84-0589 ER (MCX) (C.D. Cal., filed 27 January 1984); IBM Corp. v. Eagle Computer, C84-20109 (N.D. Cal., filed 21 February 1984).

11. IBM Corp. v. Spirales Computer, No. T-904-84 (Fed. Ct. Canada 1984).

12. Digidyne Corp. v. Data General Corp., 734 F.2d 1336 (9th Cir. 1984), *cert. denied*, 105 S.Ct 3534 (July 1985).

13. For detailed description of clean-room procedures, see Davis (1984).

14. 605 F. Supp. 816 (M.D. Tenn. 1985).

15. Kewanee Oil Co. v. Bicron Corp. 416 U.S. 470, 476 (1974).

16. See, e.g., Analogic Corp. v. Data Translation, Inc., 358 N.E. 2d 804, 807 (Mass. 1976) ("a device which has been described in trade journals and placed on the market is generally open to duplication by skilled engineers").

17. Henry Hope X-Ray Prod. v. Maron Carrel, Inc., 674 F.2d 1336, 1341 (9th Cir. 1982).

18. Hubco v. Management (1983) (an infringer making unauthorized upgrades of operating-system software was found to have violated copyright because of making (a) an internal machine copy of the unauthorized software; (*b*) a print-out of the original software without authorization; and, by implication, (c) a decompiled or disassembled copy that had to have been made to first learn how to prepare the upgrade).

19. Synercom Technology v. University Computing Co., 462 F.Supp. 1003 (N.D. Tex. 1978).

20. The hypotheses of this case study are further explored, in light of recent events, in the author's book (Davidson & Davidson 1986) and in the author's article (Davidson 1986).

References

Davidson, D. 1983. "Protecting Computer Software: A Comprehensive Analysis." *Arizona State Law Journal* 1 (December): 611, 619–630, 658–669, 678–693. See Davidson's opinions with the dissent of August, Rosenbaum, and Pokotilow (ibid., pp. 704–705, 720–721).

———. 1986. "Common Law, Uncommon Software." *University of Pittsburgh Law Review* (Summer) 47: 1037–1117.

Davidson, D., and J. A. Davidson. 1986. *Advanced Legal Strategies For Buying and Selling Computers and Software: 1986 Cumulative Supplement*. New York: John Wiley and Sons.

Davis, G. G., 3d. 1984. "IBM PC's Software and Hardware Compatibility." *Computer Lawyer* 1 (July): 11–17.

Fishman, K. D. 1982. *The Computer Establishment*. New York: McGraw-Hill, p. 28.

8. Innovation and Competition: Conflicts over Intellectual Property Rights in New Technologies

Pamela Samuelson

There is no cure for progress.
JOHN VON NEUMANN

In the last twenty years or so, the explosive growth of new technologies—many of which are radically different from prior technologies—has created something of a crisis in intellectual peoperty law.[1] Earlier, new types of innovations were usually folded into or grafted onto existing copyright or patent statutes.[2] If innovations fell between the cracks of these two systems, they would be assumed to be unprotectable by the law (Samuelson 1985).[3] In the 1970s and 1980s, however, Congress created new intellectual property systems for certain categories of innovations—most notably the Plant Variety Protection Act[4] and the Semiconductor Chip Protection Act.[5] In addition, copyright and patent law have been made to take in subject matters of such a radically different character than those for which they were designed—machine-readable software in the case of copyright[6] and genetically engineered life-forms in the case of patents[7]—that the coherence and integrity of those legal systems may be in jeopardy.[8]

The greatest challenge for intellectual property theorists in the future will be in understanding how the coherence of the existing intellectual property schemes can be maintained or a new coherence created, while at the same time finding ways for these schemes to adapt to new developments.[9] Another considerable challenge will be to assist in the accommodation of the interests of innovators and of the public at large.[10]

The first section of this chapter describes the tension that exists in intellectual property law between the interests of innovators and the interests of competitors and how this tension affects the wider public interest. The ethos that at least some innovators would have the law reflect—that they are entitled to be rewarded for the value of every use made of their works, and that others not be allowed to take any free rides by imitating the innovation— is sometimes in conflict with another ethos to which the public may subscribe—that private noncommercial copying of protected works should not be deemed infringements and that the public interest is sometimes served by some competitor copying of innovations. Both ethical and legal complications arise from this clash of ethos against ethos.

The second section discusses how this conflict has evidenced itself in the debate over what scope of intellectual property protection ought to be accorded to computer software and semiconductor chip designs. This section highlights two issues of the debate: to what extent should the overall design (or logic, or structure) of functional work of this sort be protectable, and to what extent (if at all) ought reverse engineering[11] to be permitted.[12] Although some innovative firms would urge that there be a prohibition on any reuse of software designs and any reverse engineering, such strict rules are undesirable for a variety of doctrinal and policy reasons.

The third section explores some issues that arise when intellectual property protection is extended to genetically engineered life-forms. The section responds to the chapter on the Plant Variety Protection Act by Frederick Buttel and Jill Belsky (this volume), which argues that granting a set of exclusive rights to breeders or discoverers of new plant varieties was unnecessary (because research expenditures have not risen more than historical trends would indicate was to be expected) and unwise (because private breeders will tend not to invest in the most socially desirable directions).

The section also discusses the ideology underlying the intellectual property laws—that it is necessary to create a temporary monopoly for innovations to induce creators to create—and the dangers that may attend both subscribing to it and attacking it. The truth of the ideology may be less important than that people believe in it. Even the most rational attack on such legislation as the Plant Variety Protection Act is likely to be swept aside until the faith that underlies the ideology is shaken. At present, it would seem that faith is strong, and no repeal of the act can be expected. Yet there is still an important role for critics like Buttel and Belsky to play in questioning the system's assumptions.

The fourth section considers whether academics, who presumably need intellectual property incentives far less than does private industry, should be permitted to patent the results of their research. This section responds to Charles Weiner's chapter (this volume), which argues that academic patenting is (or should be) unethical because it conflicts with the basic scientific commitment to free and open exchange of information in the public interest and that academic patenting may distort research in a harmful way. The section supports the right of academics to patent results of their research, and at the same time it recognizes that the lure of patenting may be for academic researchers as dangerous to their work as the songs of the Sirens were to ancient mariners.

Tensions between the Ethics of Innovation and Competition

What makes the copying of someone else's innovation an interesting and somewhat difficult dilemma is that it sets up an ethical conflict. There is, on the one hand, an ethic of innovation that says that copying someone else's work and then selling it as one's own (perhaps not even attributing to the innovator some credit for the innovation) is a heartless, unfair, and wrong thing to do. It is called plagiarism and regarded as a kind of theft. In some instances, the law agrees with this assessment.[13] On the other hand, there is an ethic of competition that has historically favored allowing manufacturers to produce imitations of successful products. The public interest is served when there are many competitors producing desirable products, for prices are likely to be lower.

The Sears-Roebuck v. Stiffel case[14] illustrates this conflict nicely. Stiffel was the first firm to think of making "pole lamps."[15] Stiffel sought and obtained patents both on the design and mechanical aspects of the pole lamp, hoping thereby to become the exclusive manufacturer of such lamps.[16] Sears, ever an earnest competitor, deliberately copied the design of the pole lamps, even though it knew of the Stiffel patents. Sears produced virtually identical pole lamps and sold them to the public in competition with Stiffel, and did so for a lower price. Stiffel sued Sears for patent infringement and for unfair competition.[17]

Was Sears' competitive stratagem unethical conduct? The willfulness of the infringement could not be more blatant. Yet the law and the competitive ethic that it reflects would regard the Sears imitation not merely as legal, but as a "good" thing. It was in the public interest for Sears to challenge the Stiffel patents, given that they were invalid for want of invention.[18] Stiffel's lamps, in other words, were not enough of an improvement, and were not the kind of improvement, over the prior art of lampmaking to warrant patent protection.

In striking down Stiffel's unfair competition claims, which were based on imitative copying of the now-unpatented pole lamps, the U.S. Supreme Court interpreted the patent system as intended to put into the public domain and open for unlimited copying all subject matters that were of a patentable sort but that failed to satisfy the invention standard of patentability. "Sharing in the good will of an article unprotected by patent," the Court wrote, ". . . is the exercise of a right possessed by all—and in the free exercise of which the consuming public is deeply interested" (376 U.S. at 231). Thus, under the law and under the ethic of competition, Sears' copying was a good thing.

Competition was not always so highly valued. In the Middle Ages, copying a guild's product and engaging in competition with the guild were (to say the least) viewed with disfavor.[19] During the guild era, continuity was more highly valued than was innovation.[20] Apprentices were taught to make the same thing in the same way; innovation was considered somewhat dangerous. Had the question been posed, both innovation and competitive copying would probably have been regarded as unethical as well as unlawful. As the economic systems underlying Western culture changed, competition and innovation became more valued (Renard 1918, pp. 116–120). Although public confidence in capitalism has occasionally waned,[21] the vast majority of the American public would seem to be satisfied with it.

Competition is a particularly strange ethic. If done with competition as a motive, certain acts are considered good and lawful; if done with no competitive aim, they may be seen as tortious and unethical. Tuttle v. Buck,[22] a case from the turn of the century, illustrates this point. Tuttle was a barber, and Buck, a banker. For a long time, Tuttle was the only barber in the town of Howard Lake, Minnesota. Buck became so displeased with Tuttle's barbering that Buck set up a competitive barbershop. Tuttle sued Buck for damages, claiming that Buck had set up this competitive barbershop solely for the purpose of running Tuttle out of business. Buck's response was that even if what Tuttle said were true, what he did was not wrongful. The court decided that if Buck had set up the barbershop with a genuine competitive intent, then even if he wanted to run Tuttle out of business, such an action would not be bad but good. But if Buck just wanted to run Tuttle out of business and had no genuine competitive intent, then Buck would have acted wrongfully, and Tuttle could recover for the damage to his business.

The antitrust laws[23] reflect the same faith in the essential goodness of competition as Tuttle v. Buck, but with an interesting difference. These laws reflect a judgment that if there are no checks on what one firm can do to another firm in the name of competition, then at some point competition itself may be the victim of the competitive process, to the detriment of the public.[24] The antitrust laws provide a set of checks on competitive behavior. They label certain types of competitive acts—those which threaten competition—as anticompetitive. Engaging in anticompetitive conduct is a crime as well as a civil wrong.[25]

The history of antitrust enforcement shows that it is often hard to distinguish between "hard competition" (which the ethic of competition regards as "good") and "anticompetitive conduct" (which is an "evil").[26] In a price war between two firms, for example, one firm may be attempting to drive the other out of business. The winning firm would then dominate the market and charge excessive prices. When this results, a price war may be labeled as "predatory" and "anticompetitive." It is also possible—indeed much more likely—that a price war is simply the "free-enterprise system" at work bringing the most and best products to the public at the lowest possible cost.[27] The antitrust law reflects a judgment that this kind of conduct must be supported.

Antitrust law has long had an uneasy relationship with intellectual property law.[28] Intellectual property law creates monopolies of a kind. Antitrust law views monopolies with great suspicion. That some copyright owners and patentees have acted in a "predatory" manner has tended to reinforce the negative image of monopolies.[29]

The Antitrust Division of the Justice Department has an attitude of reconciliation toward intellectual property and antitrust law.[30] That is, the Reagan administration tended to stress the compatibility of intellectual property "monopolies" with the basic tenets of the antitrust laws. There only *appears* to be a conflict, say the reconciliators. Intellectual property law creates incentives to innovate, they argue, and innovations stimulate competition, perhaps awakening otherwise noninnovative firms to become innovative to stay in business.

Whether this attitude of reconciliation will persist, or whether the more populist antitrust attitude toward intellectual property will reassert itself, is hard to predict. Probably much depends on how patentees and other intellectual property owners comport themselves in the coming decades. Absent some blatantly unfair and well-publicized predatory conduct by patentees, most people may simply accept whatever prices and terms intellectual property owners offer for their goods.

If the new attitude of reconciliation persists, competition as an ethic will likely also persist, but perhaps in somewhat altered form. Copying of the sort involved in the Sears case may gradually shift from being a legitimate, ethical, and desirable competitive act to an illegitimate, unethical, and undesirable act, at least in all instances where it would be possible to produce a competitive product dissimilar in design to the product first brought to the market by an "innovator."

Software and Semiconductor Chips: Misfits in the Copyright System

The difficulties of accommodating the conflicting interests of innovators and their competitors are epitomized in recent attempts to find appropriate forms of intellectual property protection for two economically significant technological advances: machine-readable computer software and semiconductor chips. In both cases, the initial legislative proposal was to amend the copyright statute to make these technologies protectable under that law.[31]

Certain features of copyright were attractive to both industries: a lower standard of originality than required in patent law,[32] the ability to protect overall designs (compared with the need in patent law to be quite specific about inventiveness of specific features),[33] and the existence of international treaties and conventions concerning copyrighted subject matters (U.S. Senate 1984).

One serious doctrinal problem with copyright protection for both technologies—their utilitarian character—had not arisen when software was under legislative consideration, but emerged as an insurmountable obstacle in the debate over protecting semiconductor chip designs. Copyright has traditionally considered "utilitarian" works—those having a function beyond merely conveying information and displaying an appearance—to be unprotectable.[34]

Machine hardware, which clearly has functions in addition to its visual or informational character, is not copyrightable. Semiconductors, which were hardware, therefore, could only be made copyrightable by dramatically changing the character of the copyright system, which Congress decided it was not prepared to do (Samuelson 1985, pp. 471, 501–506). Instead, a brand new form of intellectual property protection was devised for semiconductor chip designs (Kastenmeir and Remington 1985). Machine-readable forms of software also perform machine functions and merely substitute for hardware components (Pratt 1984). Thus such software also has a utilitarian character. Congress and the national commission on software seemed oblivious to this utilitarian character, however, and hence to the consequences of making software one of the subject matters of copyright. So today, copyright protects software (Samuelson 1984).

The industry and its lawyers are now trying to define what copyright protection for software means. Innovators, competitors, and users have somewhat different views on this subject.

Some segments of the U.S. software industry and their lawyers are seeking more than protection against those who exactly duplicate and distribute for profit the machine-readable version of their programs (who they describe as pirates). They also want the intellectual property system to protect them against both those who "steal" the logic and structure or overall design of their programs, and those who reverse engineer the machine-readable code to discern the pattern of instructions that make up the program.

To those who would brand reuse of software designs and reverse engineering as illegal and unethical, the motivation for the action—whether competitive copying, scientific curiosity, or simply user frustration over a bug in the code that was interfering with its operability—is irrelevant. Those who would brand design copyists and reverse engineers as pirates justify their exaltation of the ethic of innovation over the ethic of competition by insisting that as long as one can develop a piece of software by means other than copying someone's design or reverse engineering, the interests of competition are adequately served (Davidson 1986c).

There are longstanding doctrinal and policy reasons for rejecting the theories that reuse of software designs (that is, reuse of their logic and structure) and reverse engineering of software should be unlawful under copyright law. Yet courts in numerous software copyright cases have sidled away from applying traditional limitations on the scope of copyright protection,[35] almost as if they fear that if copyright protection for software is not made strong, one of the premier hopes for the future economy of the United States will be in jeopardy. The serious consequences of abandoning traditional doctrines of copyright—desertions that threaten to destroy or radically alter the integrity of the system—seem to have escaped all but a few of those who have addressed the question.[36]

Software designs are essentially engineering designs (Spector and Gifford 1986). The traditional rule of copyright holds that engineering drawings (that is, written depictions of engineering designs) can be copyrighted, but the engineering design itself (that is, the organization of the functional elements depicted in the drawing) is not within the scope of the copyright (Katz 1954). It is considered "the idea," the "system," or the "practical art" that the drawing reveals. The "idea" of the engineering design was considered to be in the public domain when the work is "published."[37] This principle was codified in the 1976 revision to the U.S. copyright law (17 U.S.C. secs. 101, 113 [b]).

Despite the long tradition in copyright of not protecting engineering designs and engineered products manufactured from them, and despite the statutory codification of the principle, courts in some recent software copyright cases have seemed to extend copyright protection to the "design" or "logic and structure" of programs.[38] The Office of Technology Assessment's study, *Intellectual Property Rights in an Age of Electronics and Information,* observes that "in theory, none of these rulings is permitted under traditional copyright principles. This is not because the courts have misinterpreted copyright law, but because *copyright law cannot be successfully applied to computer programs*" (1986, p. 81). The justification for this opinion is the utilitarian character of software. Utilitarian subject matters simply cannot be accommodated into a system that has as an integral principle the nonprotection of functional or utilitarian works.

What is most worrisome about the push by software industry lawyers to remake the copyright law in their own image is that it will destroy the integrity of the system that has served societal interests quite well for the past several hundred years (Office of Technology Assessment 1986, pp. 3–15). During hearings concerning whether semiconductor chips should be protected under the copyright or a sui generis system, Ray Patterson of Emory University School of Law urged Congress to choose the latter route:

> The ultimate issue is the problem of integrity in the law of copyright. By integrity, I mean consistency in the principles which the law encompasses. While consistency for its own sake is a virtue of small consequence, consistent principles for a body of law are essential for integrity in the interpretation and administration of that law. (U.S. Senate 1983)

Congress did choose a sui generis form of protection for semiconductors, thereby preserving the principle of nonprotection of utilitarian designs, but this message seems to be slow in reaching the courts in software cases.

For the software industry, much is at stake in the matter of reusability of software and its designs.[39] If there are only a few efficient ways to structure the organization of instructions to perform a particular task, the first few companies to write such a code may "lock" up the efficient design for seventy-five years or more.[40] The impact this restriction may have on competition and on the rate of innovation in the software development business may be negative.

Those who believe that reverse engineering is or should be unlawful argue two points: first, making a copy of copyrighted software for reverse engineering purposes infringes the copyright; and second, it constitutes a misappropriation of the trade secrets (that is, the ideas) that the software may contain.[41] But the strict rule against reverse engineering that some industry lawyers are arguing for is not only unenforceable; it also runs counter to existing practice. Computer scientists, hackers, and many software engineers think that nothing is wrong with decoding software to study the design and logic of it. Although one software case in 1985 took notice that reverse engineering of machine-readable code is a standard practice in the software industry, at least some other cases have taken the strict infringement approach.[42]

Hubco Data Products Corp. v. Management Assistance, Inc., is a case that illustrates the strict approach.[43] Management Assistance, Inc. (MAI) produced an operating system for use in conjunction with its computer. MAI marketed several versions of this operating-system software; the versions differed from one another only as to the presence or absence of *governors* (or blockages) that MAI had installed to restrict the memory capacity and certain other features of the operating system. The least expensive version of the MAI system was heavily governed; the most expensive was ungoverned.

Hubco Data Products obtained copies of the different versions of MAI's software (apparently in a lawful manner) and caused printouts to be made of the 1's and 0's constituting the machine-readable form of the different versions of the MAI programs. By comparing the patterns of the different versions, Hubco was able to discern where the governors were located in the cheaper versions and what the pattern of 1's and 0's had to look like for the governors to be removed. Hubco then contacted MAI's customers and asked those who had purchased rights to a cheaper version whether they would like to have Hubco come in and fix the MAI software to make it work like the more expensive version. After a time, Hubco developed a program that would automatically remove the governors when run on an MAI customer's system.

Not surprisingly, MAI objected to Hubco's entrepreneurial activity and charged Hubco with infringing its copyright and misappropriating its trade secrets. The court held that although the trade secrets had probably been lawfully reverse engineered, Hubco had infringed MAI's software copyright. Hubco was enjoined from using its governor-removal program, as well as from making on-site visits to customers of MAI for the purpose of removal of the governors.[44]

It was easier for the court to justify its conclusion that the Hubco program infringed MAI's copyright than it was to justify the ban on the on-site visits.

Hubco's governor-removal program consisted largely of a copy of the upgraded MAI program. The basis for the infringement claim relating to the on-site visits was that Hubco had made printouts of the upgraded versions of the MAI program. Hubco brought those printouts to the MAI customer's place of business and used them as a basis for making changes in the MAI software. These printouts were, the court found, infringing copies of MAI's program.

The court did not discuss that it might have been necessary to make such a printout in order to reverse engineer the MAI program. While the court might understandably be very concerned about Hubco's development of a business at the expense of MAI, the Hubco decision has serious implications for all owners of copies of software. If this case became the standard by which all software reverse engineering was judged, then neither a customer nor a competitor would ever be able lawfully to reverse engineer a copyrighted piece of software. Yet there are many situations in which user reverse engineering is probably justifiable, as the following story illustrates.

Assume that a private U.S. firm developed an advanced avionics system for a strategically important military airplane. Assume further that the United States learned that the Soviet Union had discovered how to exploit a weakness in the avionic software aboard the plane. The owner of this proprietary software could not provide the changes in less than a year and the Air Force had inadequate documentation to fix the software itself. Assume then an Air Force employee was able to take on, successfully, the task of reverse engineering and to fix the problem in only a month. Was that employee a pirate? Should he be held liable for copyright infringement? Or was he acting in a manner that the law should excuse despite a technical violation of the copyright when a copy was made of the software for reverse engineering purposes?[45]

In this hypothetical example, the person who did the reverse engineering might have a quite different perspective on whether it is ethical or legal than will the software producer and its lawyers. The copyright system cannot easily resolve this dilemma. As the Sony Betamax case illustrates, courts will tend to defer to practices that the public regards as acceptable conduct even though the intellectual property producers protest about the "theft" to which they are being subjected.[46]

The consumer-user reverse engineer situation is one in which it will be very hard to convince the public that copyright liability should be imposed, even though the argument that this may be taking money out of the pockets of industry is basically sound. Even some competitor reverse engineering should be tolerated, for the practice is widespread throughout the software industry.[47] The traditional predominance of public policy concerns makes copyright less than ideal as a system for those who wish to reap the maximum reward for innovations.[48]

The semiconductor chip design protection statute of 1984 reflects an appropriately permissive attitude toward reverse engineering of chip designs, a permissive attitude that the chip industry itself strongly supported.[49] It does not infringe a chip designer's rights under that law for either a competitor or anyone else (such as a computer science professor) to make a copy of a chip design for reverse engineering purposes (17 U.S.C. sec. 906). Nor is it an infringement for a competitor to use that copy in the process of preparing another chip design (Raskind 1985, pp. 403–411). This is so despite the fact that those who reverse engineer a chip may have taken something of a free ride on the innovation of another company.[50] Perhaps courts in copyright cases raising software reverse engineering issues ought to take a similarly permissive attitude.

Life-Forms as Intellectual Property

Software and semiconductor chips are not the only new technologies to present the intellectual property system with some difficult problems. Genetically engineered life-forms have also been testing the boundaries of the intellectual property system.[51] The more sophisticated genetic engineering becomes, the more serious the problems may become.

To most people, there probably is something very peculiar about the idea of granting patent rights in life-forms to any human being. Yet, through the Plant Patent Statute (35 U.S.C. secs. 161–164), the Plant Variety Protection Act (7 U.S.C. secs. 2321–2582), perhaps the Utility Patent,[52] and the Copyright Statutes,[53] the peculiarity of giving humans exclusive rights over life-forms is becoming commonplace. Tweaking a gene, in order to learn what mutation might then occur and in the hope of finding a commercially marketable mutation, certainly seems a different kind of inventive process than designing new hammers, plows, or airplane engines.[54] It is far from clear that legal rules applicable to new machines can be applied to new plants or other living things (*Wisconsin Law Review* 1984; Allyn 1933).

Congress recognized this in 1930 when it passed the Plant Patent Act for asexually produced plants. This is a kind of sui generis statute that was grafted onto the General Patent Law (35 U.S.C. secs. 161–164), although the two laws differ from each other in a number of respects.[55] More recently, Congress passed the Plant Variety Protection Act, which creates a patentlike protection for sexually reproducing plants.[56] This act is explicitly a sui generis system, radically differing from the patent statute (having, for example, no requirement of inventiveness for new plant varieties)[57] and yet sharing its underlying rationale, the need to grant exclusive rights to induce investments in this form of innovation (U.S. Senate 1970).

In their chapter (this volume), Buttel and Belsky question the advisability of granting patent or other intellectual property rights in plant-related innovations. They point out that the premise underlying enactment of the Plant Variety Protection Act—that it would increase the level of research and development efforts in that field—has not been borne out. They report that although private research and development in the field has increased, the increase is no more than suggested by historical trends without the act.

This is surely not the only time that Congress has passed a law that did not have the intended effect. And it will surely not be the only time that a law passed with an intent unachieved will remain on the books. Buttel and Belsky may be correct; the level of investment in plant research and development may not have been affected by passage of the act. For the moment, however, it does not matter. Neither the Plant Variety Protection Act nor any of the other forms of intellectual property laws will be abolished or modified simply because they have not lived up to their promise.[58] Faith is not so easily shaken.

To say that it is on faith that most people believe that granting exclusive rights will lead to high levels of innovation is not to say that it is not true. What matters is that most people in this society believe it, and that this faith guides people's actions.

The modern faith in intellectual property does not seem likely to be shaken soon, primarily because that faith is supported by such evidence as high levels of innovation, high levels of investment in innovation, and the concomitant prosperity. The intellectual property laws may not have been responsible, but most observers believe that those laws have played a part. If intellectual property law is stifling research or steering research in the wrong direction, our collective faith may keep us from recognizing it.

There is a second prevailing national faith to which Buttel and Belsky seem not to subscribe: that private enterprise, not government, should sponsor research aimed at new product development. From the standpoint of the free-market economist, it was appropriate for the government in the nineteenth and early twentieth centuries to support research and development aimed at improving, producing, and distributing seeds and other plant products because the private sector was too deconcentrated to support such research efforts. Now, however, the market is less deconcentrated, and it has become economically feasible for an element of the private sector (the seed industry supported by multinational parent companies) to take over that function. From this perspective, the public research institutions should, therefore, be abolished or take on other functions that the private sector cannot or will not perform, such as basic research in biogenetics.

The free-market theorist would point out that Buttel and Belsky themselves admit that competition in the seed industry is vigorous, the market is not unduly concentrated, and monopoly profits are not being reaped. Moreover, the passage of the Plant Variety Protection Act does seem to have resulted in the production and distribution of many new plant varieties, which Buttel and Belsky dismiss as product differentiation (that is, not truly new products, but only slight variations on preexisting products). To the free-market theorist, the situation is good.

Buttel and Belsky, however, do not think it is good. Woven through their argument are populist antitrust notions about what corporate conduct is desirable and what is not. Take, for example, their apparent thesis that the private seed industry is seeking to reduce or eliminate competition from farmers, public agricultural research groups, and other companies, and that gaining intellectual property rights is their principal strategy for accomplishing this end. Buttel and Belsky also assert that the Plant Variety Protection Act has been more effective in protecting marketing investments than it was in stimulating and protecting breeding investment and that its most important impact was the "massive acquisition-and-merger movement involving many American seed firms." They seem to be arguing that large multinational firms are seeking to eliminate public research groups as competitors of their seed company subsidiaries.

Such populist antitrust attitudes are out of fashion. The current school of antitrust theorists would respond that it is inappropriate to focus on whether one particular action may be reducing competition somewhat in a certain market. The current theorists ask whether vigorous competition still exists or is possible in the marketplace. Since Buttel and Belsky admit that vigorous competition exists, these theorists would dismiss their argument.

Buttel and Belsky seem primarily concerned with the potential "evils" caused by the redirection of public agricultural research away from applied research and into basic research. Certain socially desirable types of research will not be done if the current redirection of public research efforts persists. Buttel and Belsky argue, for example, that it may be feasible to develop pest-resistant varieties of crops, and they suggest that only public research groups would conduct such research because multinational companies will only want to develop plant varieties that will be compatible with their pesticides. Farmers would certainly prefer pest-resistant crops to large quantities of pesticides. One might argue that a certain percentage of public research funds allocated to the task of developing pest-resistant plants might eventually reduce the cost of consumer food products.

I find it hard to accept this "efficiency" argument because it requires assumptions about the costs and feasibilities of developing pest-free crops, about whether private institutions would act if the existing public research institutions did not, and about how multinational companies should spend their money.

Current legislative judgments have been to direct public funding toward basic research areas in order to create a technological base for private innovation. One can argue the merits of this decision, but one need not posit a conspiracy among multinational companies to explain the choice. Faith in the private sector's ability to produce beneficial innovations is strong at the moment. If biotechnology process patents are too broad or reflect only a slight advance over the prior art, it can be expected, as Buttel and Belsky concede, that the patents "will not be respected or will be challenged in the courts." If greater abuses—such as active suppression of new technologies that may eliminate the need for the multinationals' products—occur, there may be other responses, such as antitrust actions or even the repeal of existing statutes.[59]

Patenting by Academics: Does It Distort Research?

Reliance on the power of a patent or patentlike monopoly to persuade the private sector to innovate is not the only possible solution to the perceived problem of keeping research and development expenditures—and hence the pace of innovation—at a high level. One available alternative would be for Congress to subsidize research efforts in academic communities, the fruits of which could then be made available to all who had need of them on an unrestricted basis. To a certain extent, this alternative is already being utilized, and it is fair to say that without heavy subsidization of university-based research in the computer science and biogenetic fields, the state of the art would be much less advanced than it currently is.[60] There would, indeed, be much less for private firms to develop commercially if university researchers had avoided computers and genetics as research fields. Private enterprise is, in actuality, not the only institution that the U.S. economy relies upon to feed its hunger for high levels of innovation.

One of the questions that inevitably arises when both the private and public sectors perform such research is whether academic researchers should be permitted to patent the results of their research (Weiner this volume; Blumenthal et al. 1986b). On the other hand, one could argue that the system assumes that only private profit-making firms need the incentives of exclusivity; academic researchers do not, for a variety of reasons, seem to require such incentives. Their salaries, equipment, and the like need not be recouped as similar industry expenses must be. And the universities should uphold their tradition of free and open exchange of ideas resulting from research, even commercially valuable ones (Nelkin 1984, pp. 9–30). Academic researchers are commonly perceived to be freed from the "crass" financial motivations that may characterize their industry counterparts.

On the other hand, one could argue that the patent statute does not make the right to apply for a patent conditional on working for private industry. The patent statute provides that "whoever invents or discovers any new and useful process, machine, manufacture, or composition of matter . . . may obtain a patent therefore" (35 U.S.C. sec. 101). Yet, as Weiner's chapter (this volume) illustrates, during this century there has been considerable controversy over academic patenting, a controversy that has recently been re-ignited in the biomedical research field.

Many academics object that patenting of university research results conflicts with the values of academe. Moreover, they argue, such patenting has the potential to distort the research process causing people to engage in research not because of its intrinsic importance, but because of its potential to supplement their incomes (Blumenthal et al. 1986a).

Sometimes, as Weiner indicates, the academic values favoring "pure" motives in research are so strong that it has become a matter of ethical precept that academics *not* patent the results of their research. Nowadays, most universities actively encourage faculty members to patent their work, and do so in part because the universities themselves often claim rights to a portion of the royalties (Lachs 1984). Indeed, the academic mores of the early twentieth century, as described by Weiner, seem quaint and old-fashioned.

Weiner raises several important questions. Is there a conflict with academic values when academics patent their research? Does the prospect of being able to patent research distort the academic research process? Is it necessarily a bad thing if such a conflict exists and if research does get somewhat distorted?

Is it really at odds with academic altruism about pure research for a scholar to patent an invention and then to limit its use in order to maximize the profit from those already licensed? Is it unethical for an academic researcher to develop a trade secret and deliberately withhold it from the public to ensure its commercial value? It would be hard to say that no conflict exists. It would be nice if every academic generously dedicated all of his or her valuable ideas to the public domain. Many, of course, do and have no second thoughts. But even if there is a conflict, is it appropriate to coerce such generosity by ostracizing those who choose to patent?

I would tolerate academic patenting, just as I would tolerate letting academics collect royalties on the books they write (and which few seem to regard as a potential source of distortion). Academics who have every legal right to patent their works should be able to exercise those rights. Perhaps an occasional dose of self-interest will lead to an occasional burst of altruism. This outcome seems to me as likely as that it will lead to a life of venal self-absorption.

In the 1980s many different temptations can lead an academic astray from the route of pure research and from the ideal of free and widespread dissemination of results. Weiner's study seems most valuable in the moral lessons it teaches from past experiences, where high-minded, well-intentioned academic patentees have lost their idealism and ability to be creative researchers as they fought to preserve their control over patents.

Allowing academics to patent and to do research along these lines may keep in the academic setting some very good people who would otherwise be working in industry. If enabling faculty to obtain patent rights allows the universities to retain some people, that would seem all to the good. That universities too may benefit from receiving royalties, which may support universities in an era when governmental subsidies are in decline, may also be a reason to allow academics to patent their work (Blumenthal et al. 1986a, pp. 1363–1364). Moreover, patenting products or processes may be either necessary or very helpful to get the invention to be widely disseminated (Adelman 1982).

In assessing how tempting a lure patenting is for academics, it is worth considering a number of factors. Obtaining a patent is an uncertain, tedious, expensive, and frustrating experience, and collecting the rewards from patents even more so. It may often be easier for an academic to make his or her fortune from the lottery than from the patent system (Davidson 1983). That universities may take the lion's share of any royalties that might be earned is another factor diminishing the lure (Lachs 1984, p. 263).

It is also worth considering what other temptations may bring about distortion of research. One can argue that the entire granting process—the giving of grants by public or private institutions to faculty members—has substantial distorting effect on faculty decisions about what to research. Universities have also recently been much more receptive to allowing faculty to contract with private firms to engage in research with the use of university facilities (Nelkin 1984, pp. 18–30). Much of this research is intended to be held as company trade secrets, and a faculty member may be severely constrained in his or her ability to publish the results of the research. When innovations resulting from such research are patented, at least there is some disclosure of the innovation to a wider public than simply the company that paid for the research.[61] Larger national priorities in research may also have a substantial distorting effect. The controversy over military-funded research in general and "Star Wars" research in particular draws on this argument (Melman 1974).

Despite the fact that ethical standards may have loosened to the point that patenting is now generally regarded as acceptable, Weiner is probably right to think that the issue may be re-ignited as a serious controversy with strong ethical dimensions in the field of medical advances. Medical research expenditures are at an all-time high.[62] Much of this research is done by private firms who often employ university faculty and researchers as consultants or who give grants to academics to do the research (Blumenthal et al. 1986a, p. 1362).

Private firms in these markets will be even more likely than universities or individual faculty members to regard patenting research results as desirable.

Furthermore, they will have no compunction about charging prices for the medical innovation considerably in excess of what would prevail if there was no patent. Given the risk they took and the capital they invested in research and development, the companies may feel these high prices are well justified.

It is perhaps a contradiction in this culture that Americans are often indifferent to the plight of the poor but find their heartstrings pulled at the thought of the poor being denied a chance of survival or cure that the rich can afford. Is it ethical to withhold a cure from the poor or the infirm? Put that way, it is hard to say yes. But the question can be formulated differently, as indeed it must when considering the question of whether to allow patents to issue for medical advances. This is because in the absence of patent rights, the level of investment needed to bring about the advances that will provide cures may fall off, and there may consequently be fewer advances and, overall, fewer people cured.

As the great computer scientist and mathematician John von Neumann once observed, there is no cure for progress. New technologies are presenting difficult questions that many would just as soon not address. Ethics and legal rules, however, must be adapted to new situations as they arise.

Notes

Acknowledgment: The author would like to thank her research assistant, David Lingenfelter, for his intelligent and diligent assistance in the preparation of this chapter.

1. Intellectual property law includes, among other things, the laws of patent, copyright, and trademark. Each intellectual property law applies to a specific subject matter. Copyright law, for example, is the intellectual property law for written works; patent law is the intellectual property law for machines, processes, and compositions of matter; see generally Chisum (1980). When new types of subject matter are created, sometimes they do not fit neatly into one of the preexisting categories. Therefore, something of a crisis can be created in the body of the law.

2. The Plant Patent Act was grafted on the existing patent act in 1930 by Pub. L. No. 71-245, 46 Stat. 376 (1930) (codified as amended in 35 U.S.C. secs. 161 et seq.). Sound recordings were added to the subject matter of copyright in 1971 by Pub. L. No. 92-140, 85 Stat. 391 (1971) (codified as amended in 17 U.S.C. sec. 102 [a] [7]).

3. For example, the legislative history of the Semiconductor Chip Protection Act of 1984 (17 U.S.C. secs. 901 et seq.) reviews the reasons that chip designs either fit or did not fit within the existing forms of intellectual property law. Until this act, it fell between the cracks (U.S. Senate 1983, p. 90).

4. The Plant Variety Protection Act, Pub. L. No. 91-577, Title III, 98 Stat. 3347 (codified in 7 U.S.C. chap. 57 and 28 U.S.C.), amended by Pub. L. No. 96-574, 94 Stat. 3350 (codified in 7 U.S.C. chap. 57) (1980); see also Pub. L. No. 97-164, Title I, sec. 145, 96 Stat. 45 (codified at 7 U.S.C. sec. 2461) (1982).

5. The Semiconductor Chip Protection Act of 1984, Pub. L. No. 98-620, Title III, sec. 302, 98 Stat. 3347 (codified at 17 U.S.C. secs. 901 et seq.).

6. The machinelike character of software makes it a radically different subject matter for copyright. See generally Samuelson (1984). See also notes 31–38 below and accompanying text.

7. See, e.g., Diamond v. Chakrabarty, 447 U.S. 303 (1980). See also notes 51–57 below and accompanying text.

8. For example, the Office of Technology Assessment explains that "one cannot arrive at a 'clear distinction' between idea [which is unprotected by copyright law] and expression [which is protected by copyright] in a computer program by using traditional copyright analyses" (1986, p. 83).

9. "The application of obsolete law to novel circumstances may end up skewing the policy objectives that the statute seeks to promote" (Office of Technology Assessment 1986, p. 91).

10. It is important that the rules of intellectual property law do not become too remote from the general public's sense of what is fair and equitable; see, e.g., Sony Corp. of America v. Universal City Studios, 464 U.S. 417 (1983). The manufacturer of the Betamax machine successfully raised a fair-use defense to Universal's charge of copyright infringement of its movies based on Sony's inducement to use the machine to copy movies; the court found that there was a public interest in private, noncommercial videotaping. This decision was more popular with the general public than with the traditional copyright theorists who would have thought the case should have been decided in favor of Universal City Studios; see, e.g., Nimmer (1982).

11. In general terms, reverse engineering means the process of taking something apart (for example, a machine) and deducing from the organization and makeup of the parts how the thing works. The reverse engineering of software may involve the printing out of a copy of the machine-readable code (a string of 1's and 0's) and analyzing the pattern to decipher what the instructions were and putting this into some form of human-readable text. There are also some automatic systems that can perform these functions (Grogen 1984). Semiconductor chip reverse engineering is somewhat easier. One need only expose the etched working surface of a layer of the chip, photograph it, and enlarge the photograph to reveal the components of the chip (U.S. Senate 1983, p. 84).

12. In the original version of this chapter, this section was a response to a paper by D. Davidson. For Davidson's argument against reverse engineering, see Davidson (1986a). For a more detailed discussion, see Davidson (1986b, 1986c, 1986d).

13. Copying a book, for example, may well infringe a copyright, and can even be a crime. In general, "anyone who violates any of the exclusive rights of the copyright owner . . . is an infringer of the copyright" (17 U.S.C. sec. 501 [a]). Criminal penalties, including imprisonment, fine, and forfeiture of the infringing works, apply to "any person who infringes a copyright willfully and for purposes of commercial advantage or private financial gain" (17 U.S.C. sec. 506).

14. Sears, Roebuck & Co. v. Stiffel Co., 376 U.S. 225 (1964).

15. Rather than gaining stability from a broad, heavy base, the pole lamp utilizes spring-loaded feet to brace a slender, hollow tube between floor and ceiling. Light sources are attached to the pole at suitable heights.

16. Mechanisms are protected by utility patents under 35 U.S.C. secs. 100 et seq. Nonobvious ornamental designs are protected by design patents under secs. 171 et seq.

17. The law of unfair competition originally developed from situations in which one company "passed off" its goods as those of a competitor, thereby "stealing" the competitor's sales. Some states have expanded this label to include other "dirty tricks" (Callman 1981).

18. To qualify as a patentable invention, the subject matter of the invention must be "nonobvious." The Patent Act carefully defines this term of art by requiring that a person with ordinary skill in the art of the subject matter (as it existed before the claimed invention) should not consider the invention an obvious use or extension of the art (35 U.S.C. sec. 103).

19. Competition was to be avoided since the merchant-capitalist could undercut the guild's pricing structure and ruin the members' livelihood by employing more men and working them harder. To this end, the guild's secrets were so jealously guarded that one who violated their trust would be imprisoned (Renard 1918). The medieval guilds even had a special conception of justice, namely, "the duty to produce work of a certain standard and the right to secure employment" (Black 1984).

20. This distrust of innovation and change has been cited as one of the principle causes for the collapse of the guild system (Renard 1918, pp. 107–115).

21. The opinion of Chief Justice Hughes in Appalachian Coals, Inc. v. United States, 288 U.S. 344 (1933), reflects a faith shaken by the economic disaster of the times. In that case, the Supreme Court allowed a cooperative agreement between coal producers to stand, although its purpose was to restrain trade. The coal industry was staggered by a drop in demand and competition that was, in some instances, "abnormal and destructive" to the industry.

22. Tuttle v. Buck, 107 Minn. 145, 119 N.W. 946 (1909).

23. The primary antitrust statutes are the Sherman Act and Clayton Act (15 U.S.C.). Section 1 of the Sherman Act (15 U.S.C.) declares illegal contracts, combinations, or conspiracies in restraint of trade or commerce. Section 2 of the Sherman Act (15 U.S.C.) outlaws acts that monopolize, or attempt to monopolize, trade or commerce. Section 7 of the Clayton Act (15 U.S.C. sec. 18) bars the acquisition or combination of companies when competition would be injured or a monopoly created.

24. See, e.g., United States v. Aluminum Co. of America, 148 F.2d 416, 429–430 (2d Cir. 1945). In this case, Judge Learned Hand recognized that "a single producer may be the survivor out of a group of competitors, merely by virtue of his superior skill, foresight and industry."

Even so, monopoly power presents an "opportunity for abuse," in the form of acts of monopolization that prevent competition and violate the Sherman Act. See also Brown Shoe v. United States, 370 U.S. 294, 320 (1962), which states that "taken as a whole, the legislative history illuminates the congressional concern with the protection of *competition*, not *competitors*" (emphasis in original).

25. "Every person who shall monopolize . . . shall be deemed guilty of a felony" (Sherman Act, 15 U.S.C. sec. 2).

26. In the context of the Sherman Act, "bad" conduct is indicated by: "1) the possession of monopoly power in the relevant market *and* 2) the 'purposeful' acquisition or maintenance of monopoly power" (15 U.S.C. sec. 2). See Hills (1985, sec. 1.54). How-

ever simple these criteria may seem, the definitions of "monopoly power," "relevant market," and "purposeful acquisition or maintenance" are subject to ongoing debate (Hills 1985, secs. 1.55–1.59).

27. See, e.g., Pacific Engineering & Production of Nevada v. Kerr-McGee Corp., 551 F.2d 790 (10th Cir. 1977) (price cuts by Kerr-McGee were held not to be attempts to monopolize by predatory pricing in violation of the Sherman Act [sec. 2], but were acts of vigorous, aggressive price competition).

28. Justice Rutledge, dissenting in Hartford-Empire Co. v. United States, 323 U.S. 386, 452 (1944), captured the essence of the conflict: "Basically [the patent and antitrust laws] are opposed in policy, the one granting rights of monopoly, the other forbidding monopolistic activities."

29. See, e.g., International Salt Co. v. United States, 332 U.S. 392 (1947) (International Salt Company's use of a patented machine to coerce buyers to purchase their salt was held to be an antitrust violation); and United States v. Loew's, 371 U.S. 38 (1962) (Loew's use of popular copyrighted films to force the use of a "block booking" scheme by theaters was held to violate the antitrust laws).

30. R. B. Andewelt, deputy director of operations of the Department of Justice, Antitrust Division, defines it thus:

> The current position of the Antitrust Division on intellectual property issues differs from the position taken by prior administrations in two significant ways. First, the Division is far more receptive than it has been to expanding the intellectual property protections available to creators of new technologies. Second, the Division has modified its analytical approach for evaluating the antitrust lawfulness of intellectual property licenses in a manner that will result in less antitrust interference with patent licensing. (Andewelt 1985)

31. Software was added to the copyright realm in 1980 by Pub. L. No. 96-517 sec. 10(b), 94 Stat. 3028 (codified in 17 U.S.C. secs. 101, 117). A proposal was made to make semiconductor chips one of the subject matters of copyright in 1983 (U.S. Senate 1983).

32. See, e.g., Raskind (1985) (discussing originality standards of copyright and patent).

33. A copyright protects the whole work; a patent applies only to the inventive characteristics of the work. See 17 U.S.C. sec. 106 (the exclusive rights of copyrights, defined in terms of "the copyrighted work") and 35 U.S.C. sec. 112 (requiring particularized specification of inventive features).

34. See 17 U.S.C. sec. 101 ("pictorial, graphic, and sculptural works" are protected by the copyright scheme, including "works of artistic craftsmanship insofar as their form but not their mechanical or utilitarian aspects are concerned; the design of a useful article . . . only to the extent that such design incorporates . . . features that can be identified separately from . . . the utilitarian aspects of the article." "A 'useful article' is an article having an intrinsic utilitarian function that is not merely to portray the appearance of the article or to convey information"). See also 17 U.S.C. sec. 113(b) (a copyright on the depiction of a useful article does not grant the holder "any greater or lesser rights with respect to the . . . useful article so portrayed").

35. See note 44 below, and Whelan Assoc., Inc. v. Jaslow Dental Laboratory, Inc.,

609 F. Supp. 1307 (E.D. Penn. 1985); SAS Institute, Inc. v. S&H Computer Systems, Inc., 605 F. Supp. 816 (M.D. Tenn. 1985); Hubco Data Products Corp. v. Management Assistance, Inc., 219 U.S.P.Q. 450 (D.C. Ida. 1983).

36. "The central problem of copyright law's continued accommodation to new technology lies in the indiscriminate application of the doctrine of idea and expression to three fundamentally different categories of works: works of art, works of fact, and works of function. Unless the law recognizes the inherent differences among these types of works, technology may make the boundaries of intellectual property ownership difficult or impossible to establish, and less relevant to the policy goals the law seeks to further" (Office of Technology Assessment 1986, p. 65).

37. See Baker v. Selden, 101 U.S. 99 (1879). See also 17 U.S.C. sec. 102[b].

38. See, e.g., Whelan v. Jaslow (1985). See also SAS v. S&H (1985).

39. See generally *IEEE Transactions on Software Engineering* (1984) and Silverman (1985).

40. Copyright protection endures for the life of the author, or the life of the last surviving joint author, plus fifty years (17 U.S.C. sec. 302[a] and [b]). If the work is owned by a corporation, copyright protection lasts for seventy-five years from the date of first publication or one hundred years from the year of its creation, whichever is the shorter (17 U.S.C. sec. 302[c]).

41. Trade secret protection by itself would leave a gaping hole in protection of machine-readable software because of the time-honored right to reverse engineer a work claimed to be protected by trade secret law. If it is possible to figure out how the work was made and how to make more of the same product (by examining the work, using it, or taking it apart), trade secret law considers there to be no interference with the rights of the first manufacturer. Trade secret law has traditionally given rights only where there has been some breach of confidence between the parties or some other form of improper appropriation of the secret. See American Law Institute (1939).

42. Q-Co. Industries, Inc. v. Hoffman, 625 F. Supp. 608, 228 U.S.P.Q. 555 (S.D.N.Y. 1985).

43. Hubco v. Management Assist. (1983); SAS v. S&H (1985).

44. This opinion of the court dealt with a preliminary injunction, which requires a substantial probability of success on the merits of the case, rather than final judgment in the case. Hubco suggests that the weight of the copyright law can be brought to bear against a person who purchased a copy of a computer program and then tried to figure out what it "says." In a sense, it is like saying that someone who owns a book cannot read it. This interpretation points out the illogic of stating that the copyrighted expression of the program is readable with the "aid of a machine." In another article, I have argued at length that—consistent with the historical purposes of copyright and patent law—disclosure of the contents of a computer program ought to be *required* in order to obtain copyright protection in the first place (Samuelson 1984, pp. 705–727). As things stand now, not only can one *not* get disclosure from the author, but one can be severely penalized for even trying to do it on one's own.

45. It is just as easy to imagine a problem that Union Carbide might have with the need to enhance or improve the software controlling its chemical plants. For example,

such software is used to ensure that no leakages of deadly chemicals occur at Bhopal or that, at nuclear power plants like Three Mile Island, the rate of nuclear reaction remains stable. If such control requires one to make a copy of the software in order to be able to study it, so be it.

46. Sony v. Universal, (1983).

47. Q-Co. v. Hoffman (1985).

48. "OTA found that intellectual property policy can no longer be separated from other policy concerns. Because information is, in fact, central to most activities, decisions about intellectual property law may be decisions about the distribution of wealth and social status" (Office of Technology Assessment 1986, p. 14).

49. "The industry spokespersons . . . were insistent on preserving and encouraging the industry practice of creative copyright, a practice known to them as reverse engineering" (Raskind 1985, pp. 391–392).

50. One witness from one of the leading innovators in the chip industry testified that the research and development costs for one of his company's innovative chips had been approximately $4 million. If another firm produced exact copies of that chip, its costs would be only about $100,000. The witness estimated that, had another firm reverse engineered the design, a competitive chip could be produced for about $1 million. Although reverse engineering might permit a $3 million free ride, the witness testified that his firm could live with that. Rather, his firm wanted to be protected from the exact copyist (U.S. Senate 1983, p. 75: statement of F. T. Dunlap, Intel Corp.).

51. The patent specification, drawing, and deposit requirements seem to be inappropriate for life-forms. See, e.g., *Wisconsin Law Review* (1984) (discussing the problems of deposit and disclosure when a microorganism is patented); see also Allyn (1933) (regarding the inadequacy of language to define the patented characteristic of a plant). The natural propagation features of life-forms may create serious difficulties in view of patent law's exclusive rights to control uses of patented material.

52. 35 U.S.C. secs. 101 et seq. See also Linck (1985).

53. See Kayton (1982) (arguing for recognition of copyright in such works and simultaneous copyright and patent protection).

54. It is, for example, *not* the genetic engineer who really creates the life-form; he or she can only alter characteristics of preexisting life-forms.

55. *Sui generis* means "of its own kind of class; i.e., the *only one* of its kind; peculiar" (*Black's Law Dictionary* 1979, emphasis in original). The utility patent allows only those who "invent" to apply for protection (35 U.S.C. sec. 101), whereas under the plant patent provisions, whoever "invents or discovers and asexually reproduces" new plant varieties is eligible to obtain a plant patent (sec. 161). The scope of exclusive rights is somewhat different (35 U.S.C. secs. 154, 163), and description requirements are somewhat different (secs. 112, 162).

56. See note 2, above.

57. One need only have bred or discovered and reproduced a distinct new variety to apply for protection (7 U.S.C. sec. 2401: definition of "breeder," and sec. 2402). The utility patent statute requires a showing of nonobviousness (35 U.S.C. sec. 103).

58. In 1980, during hearings on some relatively minor amendments to the Plant

Variety Protection Act, considerable opposition to the underlying act was expressed by many who testified on the bill. Some arguments similar to those raised by Buttel and Belsky were presented at that time (U.S. Senate 1980).

59. In 1623 the English Parliament passed the *Statute of Monopolies,* which forbade the king to issue letters patent except to "true inventors." This legislation was made necessary because of abuses by kings who had, for example, given exclusive rights to sell such staple items as salt, in exchange for "royalties" (Ramsey 1936).

60. See, e.g., Dickson (1984) (discussing size of federal research expenditures, estimating federal subsidies at $42 billion for 1983). When the government funds research, it typically has expectations about making the work available to others. See generally Nelkin (1984); see also Nash and Rawicz (1983).

61. The patent statute requires the patent applicant to disclose his invention as a quid pro quo for obtaining patent protection. See 35 U.S.C. sec. 112 and Kewanee Oil Co. v. Bicron Corp., 416 U.S. 470, 480–481 (1974). See also Nelkin (1984, p. 16) (in which the author questions whether patenting really averts secrecy).

62. See Nelkin (1984, p. 24) (reporting on growth of funding for biomedical research).

References

Adelman, M. 1982. "The Supreme Court, Market Structure, and Innovation." *Antitrust Bulletin* 27: 457.

Allyn, R. 1933. "Plant Patent Queries." *Journal of the Patent Office Society* 15: 180.

American Law Institute. 1939. *Restatement of Torts*. St. Paul, Minn.: American Law Institute.

Andewelt, R. B. 1985. "The Antitrust Division's Perspective on Intellectual Property Protection and Licensing: The Past, the Present, and the Future." Paper given to the American Bar Association Patent, Trademark, and Copyright Section, London, 16 July. Washington: Department of Justice.

Black, A. 1984. *Guilds and Civil Society in European Political Thought*. New York: Methuen, p. 16.

Blumenthal, D., M. Gluck, K. S. Louis, M. Stoto, and D. Wise. 1986a. "University-Industry Research Relationships in Biotechnology: Implications for the University." *Science* 232: 1361–1366.

Blumenthal, D., S. Epstein, and J. Maxwell. 1986b. "Commercializing University Research." *New England Journal of Medicine* 314: 1621–1626.

Callman, R. 1981. *Unfair Competition, Trademarks and Monopolies*. Wilmette, Ill.: Callaghan, sec. 2.01.

Chisum, D. 1980. *Intellectual Property: Copyright, Patent, and Trademark*. New York: M. Bender.

Davidson, D. 1983. "Protecting Computer Software: A Comprehensive Analysis." *Jurimetrics* 23: 339–425.

———. 1986a. "A Black Box Approach to Software Copyright Infringement." *Computer Lawyer* 3: 25.

———. 1986b. "The Whelan Decision: Missing the Middle Ground." *Computer Law Reporter* 5: 342–346.

———. 1986c. "Common Law, Uncommon Software." *University of Pittsburgh Law Review* 47: 1080–1085.

Davidson, D., and J. A. Davidson. 1986. *Advanced Legal Strategies for Buying and Selling Software: 1986 Cumulative Supplement*. New York: John Wiley and Sons, pp. 49–56.

Dickson, D. 1984. *The New Politics of Science*. New York: Pantheon, pp. 20–23.

Grogen, A. 1984. "Decompilation and Disassembly: Undoing Software Protection." *Computer Lawyer* 1: 6–7.

Hills, C. A., ed. 1985. *The Antitrust Advisor*. 3d ed. Colorado Springs: Shepard's/McGraw-Hill, sec. 1.54.

IEEE Transactions on Software Engineering. 1984. "Special Issue on Software Reusability." *IEEE Transactions on Software Engineering* SE-1: 473–609.

Kastenmeir, R., and M. Remington. 1985. "The Semiconductor Chip Act of 1984: A Swamp or Firm Ground?" *Minnesota Law Review* 70: 417.

Katz, A. 1954. "Copyright Protection for Architectural Plans, Drawings, and Designs." *Law and Contemporary Problems* 19: 224, 236.

Kayton, I. 1982. "Copyright in Living Genetically Engineered Works." *George Washington Law Review* 50: 191, 197–218.

Lachs, P. 1984. "University Patent Policy." *Journal of College and University Law* 10: 263.

Linck, N. 1985. "Patentable Subject Matter under Section 101: Are Plants Included?" *Journal of the Patent Office Society* 67: 489.

Melman, S. 1974. *The Permanent War Economy*. New York: Simon and Schuster.

Nash, R., and L. Rawicz. 1983. *Patents and Technical Data*. Washington: Government Contracts Program and George Washington University.

Nelkin, D. 1984. *Science as Intellectual Property*. New York: Macmillan.

Nimmer, M. 1982. "Copyright Liability for Audio Home Recording: Dispelling the *Betamax* Myth." *Virginia Law Review* 68: 1505.

Office of Technology Assessment. 1986. *Intellectual Property Rights in an Age of Electronics and Information*. Washington: Government Printing Office.

Pratt, T. 1984. *Programming Languages: Design and Implementation*. 2d ed. Englewood Cliffs, N.J.: Prentice-Hall, p. 19.

Ramsey, G. 1936. "The Historical Background of Patents." *Journal of the Patent Office Society* 18: 6, 8.

Raskind, L. 1985. "Reverse Engineering, Unfair Competition, and Fair Use." *Minnesota Law Review* 70: 385, 390–396.

Renard, G. F. [1918] 1968. *Guilds in the Middle Ages*. Reprint, New York: A. M. Kelley, pp. 32–40.

Samuelson, P. 1984. "CONTU Revisited: The Case against Copyright Protection for Computer Programs in Machine Readable Form." *Duke Law Journal* 4 (1984): 663–769.

———. 1985. "Creating a New Kind of Intellectual Property: Applying the Lessons of the Chip Law to Computer Programs." *Minnesota Law Review* 70: 471, 490–492, 501–506.

Silverman, B. 1985. "Software Cost and Productivity Requirements: An Analogic View." *Computer*: 86–96.

Spector, A., and D. Gifford. 1986. "A Computer Science Perspective of Bridge Design." *Communications of the Association for Computing Machinery* 29: 268, 282.

U.S. Congress. House. 1983a. *Copyright Protection for Semiconductor Chips: Hearings on H.R. 1028 before the Subcommittee on Courts, Civil Liberties, and the Administration of Justice of the House Committee on the Judiciary*. 98th Cong., 1st sess., pp. 56–57: statement of L. R. Patterson, Emory University School of Law.

U.S. Congress. Senate. 1970. *Plant Variety Protection Act: Hearings on S.3070 before the Subcommittee on Agricultural Research and General Legislation of the Committee on Agriculture and Forestry*. 91st Cong., 2d sess., pp. 47–50: statement of J. Miller.

———. 1980. *Plant Variety Protection Act: Hearings on S.23, S.1580, and S.2820 before the Subcommittee on Agricultural Research and General Legislation of the Committee on Agriculture, Nutrition, and Forestry*. 96th Cong., 2d sess., pp. 117–123: statement of C. Fowler.

———. 1983. *The Semiconductor Chip Protection Act of 1983: Hearings on S.1201 before the Subcommittee on Patents, Copyrights, and Trademarks of the Senate Committee on the Judiciary*. 98th Cong., 1st sess.

———. 1984. *S Rept. 425*. 98th Cong., 2d sess., p. 13.

Wisconsin Law Review. 1984. Note: "Patent Protection for Microbiological Processes." *Wisconsin Law Review* 4 (1984): 1679–1709.

9. Patenting Microorganisms: Threats to Openness

A. J. Lemin

The U.S. Patent and Trademark Office had long held that living matter was not patentable, because its origin was viewed as an act of God rather than of human endeavor. One still cannot obtain a patent on an organism such as a hen's egg. In 1972, however, Ananda Chakrabarty, an employee of the General Electric Company, set out to obtain a patent on a living microorganism that had been genetically manipulated to degrade hydrocarbons for use as an energy source. The patent was ultimately granted, but only after much controversy.

In brief, the patent examiner denied Chakrabarty the patent, claiming that the bacteria consisted of artificially produced mutants, which—like naturally occurring mutants—were products of nature. In contrast, the Board of Appeals concluded that the genetically engineered organism was indeed the product of human invention and therefore patentable. Yet despite this conclusion, the board ruled that the invention was not patentable on the grounds that Congress had never written patent protection for bacteria into the law. Although Congress had passed the Plant Variety Protection Act (1970) whereby plants became eligible to receive patentlike protection, it had not legislated protection for microorganisms, and therefore the courts refused to include microorganisms under the existing patent laws. In 1980 the Chakrabarty case finally went to the Supreme Court. In a five-to-four decision the Court ruled that the genetically engineered bacterium was indeed patentable, providing of course that the new organism met all the other criteria for patentability.[1]

During the same period in which the Chakrabarty case battled its way through the court system, the Board of Appeals rejected a patent on a "biologically pure culture of a microorganism."[2] In this case, the board argued that the microorganism was a product of nature, even though biologically pure cultures of the organism cannot be found in nature. Yet following the Chakrabarty decision, the Appeals Court agreed to allow patents on biologically pure cultures of microorganisms. In the majority opinion, this court ruled that, as far as patent claims are concerned, "the fact that microorganisms, as distinguished from chemical compounds, are alive is a distinction without legal significance." Thus the way was open to patenting living microorganisms and cell lines.

The significance of patent protection for genetically engineered living organisms is that it grants the inventor the right to authorize who may use the organism. This protection, as in the case of any other patented invention, enables the inventor to profit from his invention or control its use for the benefit of the general public without the threat of unauthorized use. Through the manipulation of genetic information from cells, scientists in biotechnology are seeking insights from which they may produce new and useful products. One of the more promising areas for biotechnological knowledge is in the discovery and the manufacture of pharmaceutical products. In promoting the development of new and better drugs, patent protection has itself proven to be a vital factor, giving the manufacturer a limited yet assured opportunity to recoup the large investment needed to bring a new drug to the point where it can be safely administered.

One example should suffice to illustrate the significance of the need to protect cell lines. Genentec is a flourishing biotechnology company doing pioneering research on many worthwhile problems. As a result, in cooperation with academic laboratories on an international front, Genentec has succeeded in producing microorganisms that make a high-molecular-weight protein called *tissue plasminogen activator*. This protein is said to dissolve clots that seal off arteries and result in heart attacks, and its estimated pharmaceutical market value is half a billion dollars. The precautions that Genentec takes to protect its investment in this research include restricting the availability of its more successful protein-producing cell lines. It is interesting to note that very few U.S. patents cover specific genomic constructions. This fact may reflect the conservative attitude of the U.S. Patent Office toward these inventions. On the other hand, Genentec has been publishing its European patent applications, which contain claims to genomic-type constructions for the production of enhanced tissue plasminogen activator.[3]

Given the large stakes, it is not surprising that most industrial organizations and many academic institutions are working to protect their biotechnological research results, although such efforts may lead to a cascade of pragmatic and ethical problems, including restraints on the free exchange of plasmids and cell lines, delays in the publication of pioneering work in order to allow for the filing of U.S. and foreign patents, and the setting up of commercial biotechnology companies based on applications of academic basic research paid for by the federal government.

For instance, biotechnological research in U.S. universities is largely financed by the taxpayer. Therefore, it seems that the successful results of this research should somehow benefit the tax-paying public. Many universities have discovered that they have salable, federally financed, protectable technology, which can be assigned to them upon asking. This technology can then be sold to the highest bidder, and in some cases the highest bidder may well be a foreign company. Thus the U.S. taxpayer may be financing research that benefits foreign rather than domestic industry.

Describing the Patentable Material

In the aftermath of the Chakrabarty and other decisions, patents on results in biotechnology are now being issued. The problems of applying the existing statutes and rules to this area of research are proving formidable. In addition to the three standard requirements for patentability—novelty, nonobviousness, and utility—a patent specification must contain a description of the best known way to carry out the invention in order to permit someone "skilled in the art" to make use of the invention without extensive research-and-development costs. In the case of microbiological research, where someone skilled in the art simply could not make a specific living microorganism, the "enablement" requirement is met by depositing a viable sample of the organism in a public depository before the patent application is filed. This public depository then makes the culture available to interested users after the patent is issued.

The inventor could, of course, attempt to describe his invention in sufficient detail in the patent specification that the invention could be recreated by someone with sufficient skill. However, this seems to be possible only when the starting organism is readily available and the instructions for manipulating the genetic information are known with certainty. For instance, a bacterium might be engineered to contain the gene responsible for producing a high-molecular-weight protein whose exact structure is not known. In this case, instructions specifying how to make the gene for insertion into the bacterium cannot be written. The inventor must supply the gene.

Once the engineered microorganism is made available to the public, further problems appear. The nature of the genetic information carried by a cell line is such that it can be extracted as DNA. The DNA can then be examined in detail to find out which of its pieces might be useful for splicing into the DNA of other cells or organisms. The engineered cell in culture produces (that is, "expresses") the desired protein coded by the DNA gene. This process is under the control of a regulatory system involving nucleotide base sequences both in front of and behind the actual gene sequence. The patent literature describes such regulatory sequences for genes. These sequences are not generally specific for any particular gene coding of a particular protein and may be used as regulatory sequences for many different genes.

Identifying Patent Infringements

In the case where a second inventor uses small pieces of a first inventor's DNA (for example, a gene-expression promoter), it may not be clear whether the second inventor is infringing on the first inventor's patent. The solution to this depends, of course, on the claims made in the first inventor's patent. Having spliced a prior inventor's useful DNA into his genomic construction, a researcher may find that his new cells produce large quantities of valuable proteins. These proteins would probably have no direct structural relationship to the pirated DNA. The fact that they were obtained through an infringement of the first inventor's patent would be ascertained only through a complete sequencing of the DNA in the second inventor's newly engineered organism. Since this organism would not be available to the general public, there would be no immediate way of discovering the infringement. Consequently, there is no practical way to protect the inventor of the original cell line from accidental or otherwise covert infringement of his patent rights. It is entirely possible that the second inventor is unaware that he is using protected material.

This is not to say that biotechnologists quietly steal one another's genetic engineering constructions; however, there is a healthy foreign and domestic traffic in cell lines. It is easy to see how either the identity of a cell line or the knowledge that a cell line is protected may be lost after the culture has passed through several laboratories. Indeed, foreign academic and industrial organizations and their scientists have full access to deposited cell lines and microorganisms, and it is easy to understand how, in the excitement of scientific discovery, foreign researchers may forget the distant and complicated U.S. legal system.

Most universities and industrial firms recognize the need to promote and encourage the exchange of cell lines and plasmids between laboratories. In an attempt to improve the protection for their cell lines, some major universities require an industrial recipient of the cells to sign an agreement stipulating that the cell line will not be distributed to third parties, and that anything derived by the recipient from the cell line will belong to the university. It is the latter provision that leads industrial researchers to deliberate, in advance, whether it is really necessary to obtain the cell line before signing up.

"Anything derived" extends the ownership rights of the cell line owner and the university into an uncharted area of future research. Such an extension appears to be an unwarranted invasion of the rights of the recipient of the cell line. Of course, the recipient is under no absolute compulsion to acquire the cell line. In fact, industrial firms are becoming increasingly reluctant to sign over to a university the rights to their future inventions. In contrast to university policy, many industrial concerns have more lenient agreements with recipients of cell lines. Like universities, industries ask recipients not to pass the cell lines along to others. But rather than claim ownership over "anything derived," they ask that if inventions are made using the company's cell line, "to the extent that it is able, the institution [read *researcher*] will provide [the company] with appropriate recognition of [the company's] contribution, such as a first option to negotiate a license to use the invention or substance or a reduced royalty if a non-exclusive license is offered."[4]

If all cell lines that are bought, sold, exchanged, or given away could be the subjects of valid patents, then the supplier of the cell line could easily resolve issues of infringement by asserting his patent rights. Therefore, legal proceedings regarding patent infringement would settle where the rights of the supplier end. Yet this is not a practical solution, and one is left feeling that a (perhaps absurd) desire to protect only fanciful rights to property is seriously interfering with the legitimate scientific need to exchange tangible property in the form of cell lines and microorganisms.

Viewing the Alternatives

Some innovative approaches to protecting biotechnological property rights have been suggested. The treatment of cell lines and plasmids as trade secrets embodies one possibility. The advantage of trade secrecy is that it provides absolute protection, so long as the secret is maintained. Moreover, the courts will support the rights of the holder of a trade secret if the secret is divulged or discovered by improper or unfair methods. In addition, rights to the use of trade secrets may be bought, sold, and licensed, much as patents are. The problems lie in the fact that independent inventors may stumble across the secret and either publish it or—worse—take patent action, whether or not this is prudent. The latter action could perhaps force the original inventor to obtain a license to his own invention.

The idea of using trade secrecy to protect the property rights of a scientific discovery is antithetical to the basic tenets of academia. It appears that only one academic institution has proposed trade secrecy as a viable plan of action. In industry, however, trade secrets are commonplace, and restrictions on the publication and the disposition of cell lines are quite acceptable. In order to answer the question of whether trade secrecy is an appropriate choice, three classes of cell lines should be considered: 1) When the cell line or plasmid is a commodity readily available from multiple sources, there is no reason that the materials cannot be exchanged by researchers. 2) When cell lines are known to be easily constructed, they, too, should be made available for trading, unless there is some overriding reason to keep them secret. 3) In contrast, when genetic engineering has produced cell lines with new and potentially useful effects, availability could infringe upon the rights of the researcher while aiding the competition, whether industrial or academic. These cell lines may be held as trade secrets, but it appears that such cell lines are rather rare.

In order to attract and maintain a highly qualified research staff, an industry must offer to publish high-quality research papers by its present employees. In addition, it must assure the incoming worker, who is presumably steeped in the doctrine of free publication, that he will be given the opportunity to publish his future research. Biotechnology is a rapidly advancing field. Even if an individual or company takes patent action in order to protect an invention and then allows publication to take place, by the time the patent is issued two or three years later, the invention may have become obsolete. This problem with patent protection does not, in itself, justify trade secrecy: cell lines or plasmids kept as trade secrets will most likely suffer the same fate and become rapidly obsolete as well.

It has been seriously suggested that the act of manipulating a piece of the genetic code is analogous to writing a manuscript or composing a piece of music. If copyright protection is available for a work of authorship, then similar protection ought to be available for works by those who design and construct original sequences of DNA. A current trend in thought, however, is that the construction of original gene sequences is not a work of authorship. In either case, it seems that if copyright protection were available, its range would necessarily be quite narrow and easily bypassed by minor changes in the already redundant code sequence. Just as in the case of an inventor who independently discovers a trade secret, one who independently discovers a copyrighted DNA sequence would not be guilty of an infringement under the current U.S. copyright laws.

Upon reflection, it appears that the patenting of innovative parts of a genetically engineered nucleotide base sequence offers the best protection for the inventor. Assuming attention is paid to specific claims and that patent laws in biotechnology are derived from current chemical patent laws, it would appear that patents of biotechnology will reasonably control the use of parts or fragments of the genomic sequence, in addition to protecting the all-important gene itself. It remains to be tested whether minor differences in nucleotide base sequences that result in equivalent protein products will be patentable, despite original base sequence and protein claims. An appropriate analogy to this situation is the cake mix patent. While it is theoretically possible to obtain a patent on a cake mix, in order to avoid infringement one need only change, for example, the sugar content of the mixture from the necessarily limited range in the patent. The second cake will either be sweeter or less sweet than the original, but it would still be cake.

Biotechnological patent law and trade secrecy practices are still being developed. The more significant issues (which perhaps will only be settled in court) remain to be resolved. Given the potentially high stakes, especially in the area of pharmaceutical development, a reluctance is growing toward the free exchange of plasmids and cell lines. With time, this reluctance could seriously impair the traditional, free exchange of materials and information.

Notes

1. Diamond v. Chakrabarty, 447 U.S. 303 (1980).
2. Application of M. E. Bergy, 596 F.2d 952 (1979).
3. European Patent Application No. 117059.
4. From a statement published by the Monsanto Company.

10. Patenting Body Parts: A Sketch of Some Moral Issues

Carl F. Cranor

In 1976 a team of doctors in Alaska diagnosed John Moore as having hairy-cell leukemia, a rare and potentially fatal form of leukemia. Disturbed by this diagnosis, Moore went to the UCLA Medical Center for a second opinion. There, doctors confirmed that he did indeed suffer from hairy-cell leukemia and agreed that the only appropriate treatment for his condition would be the removal of his spleen. The splenectomy was successful and Moore's recovery seemed remarkable for one with such a threatening disease. Subsequent blood samples, presumably used to monitor his condition, showed that his progress was good, even miraculous. Over a period of seven years, the doctors at UCLA asked that he return to the medical center for blood withdrawals; at each visit, he was told that his progress was excellent.

During one of these visits to the medical center, Moore's physician, Dr. David W. Golde, told Moore that he had certain "unique characteristics" that were of interest to the research physicians. Whenever Moore asked about their research, he was told only that it focused on the nature and cause of his and other diseases and concerned "the betterment of humanity."[1] During one of his last visits, he was presented with a consent form that would grant to the university "any and all rights I, or my heirs may have in any cell line or any other potential product which might be developed from the blood and/or bone marrow obtained from me."[2] During his final visit, he was again asked to sign this consent form, but he refused. When he again requested information about the ongoing research and possible commercial interests, he again was given only vague replies. Later, upon discovering that Moore had not signed the waiver, Golde telephoned Moore and sent waiver forms to Moore's home, urging him to sign. Once again, Moore refused and, at that point, sought legal counsel.

Subsequently, Moore filed a legal complaint against his doctors and the University of California regents with thirteen counts alleging harm through the breach of implied contractual relationships and the breach of tort law duties.[3] During discovery prior to trial, Moore and his lawyers claimed to have found that Dr. Golde had filed for a patent on "unique T-lymphocyte line and products derived therefrom," based entirely on cells derived from Moore's body.[4] In addition, they claimed that Golde had entered into an agreement with a biotechnology company for half a million dollars to develop commercial products from Moore's blood cells and their products.

Moore's case raises a number of issues that may clarify important concerns regarding property rights in body tissues. First, should a person have property rights in various body tissues, or (as in Moore's case) should he have property rights in unique or at least very rare body tissues, such as blood products? Second, which property rights should one have in one's tissues; in particular, should one be permitted to sell one's body tissues or merely to give or bequeath them to others? Third, under what conditions, if any, should one be permitted to exercise one's property rights? This paper assesses aspects of these questions.

Preliminary Issues

John Moore claims that his blood samples have unique characteristics, that they have been patented by Dr. Golde, and that they are potentially quite valuable in the marketplace of biotechnological products, as evidenced by the sale of the patent rights to a biotechnological firm. In addition, he claims that his doctor was profiting from these unique body products and the patents or products derived from them. Moore's legal suit and oral testimony make it evident that he believes he holds property rights over his "unique" body tissues.[5] Yet, according to some observers at a congressional hearing concerning biotechnological developments, Moore's blood should not have been patented in the first place because it may not be a unique biological product, but one that is quite common. (Uniqueness is a requirement for patents.)[6]

It is, however, agreed that Moore's case is unusual, and for this reason the case may appear to have limited value as a legal or moral precedent. Yet even if cases such as Moore's are rare and likely to remain so, they raise interesting questions, including philosophical ones, that must be discussed. (Just prior to publication of this volume, a "three judge panel of the California Court of appeals in Los Angeles ruled that Moore had property interests in his bodily products and therefore [this issue] could go to trial" [*Los Angeles Times* 1988]). In what follows, I consider a hypothetical case similar but not identical to that of John Moore, despite my references to "John Moore's case" or to "John Moore's tissues." Whereas the actual circumstances of Moore's case are often difficult to ground in fact, the issues that it concerns provide an essential springboard for discussion. For the most part, I focus on property rights issues, not on matters of consent to the various procedures.

Moore's experience illustrates in a dramatic way the recent trend toward commercialization that has seized the biotechnological revolution. Due to a number of recent court decisions permitting patents on biotechnological creations, the wealth and income potential of these innovations have created incentives for entrepreneurs to develop related, marketable products. Permitting intellectual property rights in unique or even rare body tissues creates a system of incentives for possible entrepreneurial exploitation of the donor or supplier. One obvious concern is that such a system may embody a kind of distributive unfairness to the supplier unless the supplier receives some of the commercial benefits. The failure to grant the supplier property rights over his or her own body tissues may further intensify the supplier's feeling of resentment over this apparent injustice. An altogether different concern is that it is morally repugnant to turn a person's body (or body parts) into a commodity in the marketplace by creating salable property rights in it.

Below, I sketch some of the main moral issues that bear on the position that an individual ought to have patentable property rights in his or her unique or very rare body tissues. By contrasting this position with various alternatives, some of the moral issues inherent in this stance will become clear.

Property Rights in One's Tissues

John Moore's claim that unique bodily substances were both patented by Dr. Golde and sold for a considerable profit to a drug company makes Moore's case one of particular interest. One striking feature about the case is that it shows how biological material unique to a particular individual can be highly useful to many others. It is also interesting to note that such material can have substantial value, although it seems odd that tissues with unique properties would be valuable to thousands of others. Even if such material embodies unique properties that allow it to be patented, it does not follow that all such patented material is thereby valuable or should be patented. (For example, hundreds of thousands of biological experiments, which initially seem commercially valuable, fail each year.)

If Moore's claims are true, the commercial value to him of his tissues would stem from, first, the possibly unique character of the tissues and, second, the possibility that his doctors made very few modifications in the tissues in order to make them valuable to a drug company. The uniqueness of Moore's fluids, in addition to satisfying one condition required for patents, would make them more valuable than if they were common. However, if his doctors had greatly modified the tissues for patenting, then his claim to recoup some of the value of the end product would be considerably weakened. (This point is even more evident when hundreds of patients have donated tissues, parts of which are purified and then biologically modified to a considerable extent in creating a new product. Claims to receive payment for donations in such cases are weak or nonexistent.)

Consider the analogy to especially valuable natural products, such as diamonds and gold. Both substances derive an initial value from their relative scarcity and both can have considerable value added to them by the workings of a goldsmith, a diamond cutter, or a jeweler. To the extent that further work is added to the diamonds or gold, the original seller of these raw materials would have a lesser claim to the added value of the final product.

Several other aspects of Moore's case bear on the right to transfer his tissues. Moore's blood plasmids were found to be replenishable and were neither vital—at least in the quantities withdrawn—nor were they mere by-products of life, such as urine, feces, saliva, or expired air. In addition, his plasmids were somatic cells in contrast to germ cells. All of these characteristics strongly influence one's judgment of Moore's case. If his tissues were not replenishable, or if they were vital to his life processes in the quantities withdrawn, one might be more inclined to deny his right to transfer these materials, for by transferring nonreplenishable or vital tissues, he might be shortsightedly bringing himself harm.

Furthermore, if instead of somatic cells, his plasmids were germ cells, which have the genetic possibility of duplicating his particular genetic traits, one might decide to prohibit the transfer of them, for germ cells can develop into an organism or one of its parts, and somatic cells cannot. The transfer of germ cells might raise much more controversy, given the recent debates over test-tube babies and cloning. Finally, if Moore's tissues were mere by-products of his life processes, and were thus systematically discarded in the course of living, their value would be small and any claim to the right to benefit from the sale of such materials would also be denied.

From the above discussion, it seems plausible that one should have some rights in one's unique yet replenishable, nonvital, and somatic body tissues. However, what specific rights should one have in these tissues? Should one have property rights in such substances, or should one have some rights, but not necessarily *property* rights?[7] The standard, full, legal complement of property rights in one's body tissues would include the following: (1) the right to possess *X*; (2) the right to use *X*; (3) the right to manage *X*; (4) the right to the income, if any, from *X*; (5) the right to the capital of *X*; (6) the right to security of *X*; (7) the power of transferring *X*; (8) the absence of term with regard to *X*; (9) the prohibition of harmful use of *X*; (10) the liability to execution with regard to *X*; and (11) the residuary character of *X*.[8]

The rights to possess, to use, and to manage, in addition to the right to the security of, as well as rights 8 through 11, seem uncontroversial in this case (although I do not argue this point here). For example, it is not, I think, problematic that a person have the right to security over his or her body parts, for if others could expropriate them at will, one's life and health would indeed be insecure. In addition, it would be odd for a person to exist without the right to manage his or her body tissues. However, even though an individual has rights 1 through 3, 6, and 8 through 11 to his or her tissues, these rights may not count as property rights in an ordinary sense. The rights to the income from and to the capital of these tissues, in addition to the power of transfer might instead be considered much more central to the idea of property. Consider two controversial rights: the legal power to transfer (sell) one's tissues (and the exact form this power takes), and the right to the income from the sale of tissues.

The Right to Transfer

The right to transfer *X* is legal power—the legal ability to convey validly to others as well as the legal permission to exercise this power (Hohfeld 1923). The exact form that this power takes depends on the extent of the power. The power to give or to bequeath body tissues to others seems relatively uncontroversial, and this would certainly be true for blood fluids. Except in highly unusual circumstances, there seem to be no objections in principle to a person's choosing to donate whole blood or blood serum to others for transfusion or for research. Nor would there be much objection to a person's bequeathing such fluids after death, if that is biologically possible. However, the right to transfer one's blood fluids by sale is, perhaps, the central and most controversial issue expressed by Moore's case.

Congress and several states have decided that body organs may not be sold for transplant purposes, perhaps reflecting deeply held beliefs that there is something very frightening about a marketplace in human organs. Nonetheless, throughout the United States, one may sell whole blood or blood plasma for transfusion or research purposes, although the sale price in both cases is ordinarily quite modest. These practices have, however, failed to clarify the issue of tissue sale as a whole, and thus a further question must be raised: When should one be permitted to sell one's tissues and when should this sale be forbidden?

This question has been addressed by Thomas Murray. He argues that we think that our body tissues and organs, for the most part, should be treated as valued and precious parts of ourselves and should not be denied respect in certain ways (Murray 1987). For example, we would be very offended to find that even diseased organs that had been removed from our bodies were used to decorate a physician's trophy wall. Our interest in our organs continues after removal and any permission that is granted to others should be consistent with this view of ourselves and our tissues. The proper conception of conveying tissues to physicians or to others is best seen as a gift transaction. (Some transactions are sales, such as whole blood or plasma, but we may feel somewhat uncomfortable with these exchanges).

Gifts are made with implicit conditions. In particular, we convey gifts for certain purposes or with certain understandings, and if these are violated, we would rightly feel outraged or betrayed. For example, if we had given cherished family recipes to a friend, who then put these into a book or pamphlet in order to sell them for a profit, we would be greatly offended because our friend would have violated the conditions of our gift.

With regard to the practice of donating diseased organs or excised tissues to physicians for research, we would rightly feel betrayed if these were used for commercial purposes without our knowledge or permission. Nevertheless, times are changing, and progress in bioengineering may force us to confront the commercial rights of tissue donors. And yet, because there are so few cases in which it has been deemed appropriate for a particular individual to share in profits from his tissues, we should not yet jeopardize ongoing and valuable patient/doctor and subject/researcher relationships by commercializing them further. In addition, because of the ambiguity already characterizing these relationships, there should be no change in informed-consent forms nor any congressional action for now.

However, despite the fact that so few cases exist in which we might think that a donor is entitled to sell his tissues, and despite the ambiguity of the patient/doctor, subject/researcher, and supplier/vendor relationships, commercialization of body tissues and the assertion by individuals of property rights in their tissues pose several threats to the gift relationship between patients and medical scientists. Murray describes some of these dangers, explaining that "public trust and confidence in the scientific profession may be damaged if scientists are seen as greedy or especially as having taken advantage of a person who has made a gift from his own body to science" (Murray 1987). In order to avoid these harms, Murray ultimately recommends that representatives of science, industry, academe, and the public draft a voluntary policy as a first step, and if their efforts fail, only then should Congress step in.

Murray's argument provides significant observations regarding the present practices of tissue and organ donations as well as the potential dangers of sales in human tissues. However, by recommending that the concerned parties work out a voluntary agreement, Murray fails to resolve central issues about what a desirable practice in tissue transfer might be within the new environment created by the biotechnological revolution. Although this short chapter cannot accomodate a full discussion of such a practice, by focusing on the relationship between doctor and patient, it can advance the assessment of various practices. Consider Murray's observation that one of the current problems regarding tissue transfer is the multiply ambiguous relationship between a doctor and patient. Is this relationship one of doctor/patient, researcher/subject, or vendor/supplier? To the extent that this relationship remains ill-defined, the overall treatment of the patient will also be ill-defined.

The Validity of Sales

Moore's case suggests the particular outrage that a patient is likely to feel from unjust treatment by his or her doctor. In Moore's case, two aspects best characterize this apparent injustice: that the purpose of withdrawing Moore's blood was not made clear to him, and that the relationship between Moore and his doctor often shifted from that of doctor/patient to that of researcher/subject to that of vendor/supplier. Because Moore went into the hospital for diagnosis and, if necessary, treatment of hairy-cell leukemia, it was reasonable for him to expect that a relationship based on trust would exist between him and his doctor. Such a trust would necessitate that the doctor follow specific principles of conduct and that he be committed to treating his patient's disease and taking care of his health.

However, once Moore's doctor became interested in using him as a research subject, the doctor/patient relationship changed in subtle ways. In addition, once his doctor began the procedure to patent Moore's blood and to withdraw blood for that purpose, the relationship changed once again, this time assuming a commercial nature. The expectations, trust, and principles of conduct appropriate to the doctor/patient, and even the researcher/subject, relationship are quite different from those appropriate to the commercial relationship between vendor and source. Due to the ambiguous nature of Moore's relationship with his doctor, it is not surprising that Moore felt outraged.

Such ambiguity, in fact, obscured the intended use of Moore's blood. It is one thing for a patient to give blood to help in the diagnosis and treatment of his own disease. It is another for one to make a gift of blood to promote research that will not necessarily help in one's own treatment. It is something else entirely to be a supplier of blood to one who would patent it and develop it for commercial use by a drug or biotechnological company. If these three relationships were kept separate, problems such as Moore's, and those with which Murray is concerned, could be avoided altogether.

Their prevention might be achieved by forbidding both primary-care and research physicians from becoming commercial vendors of a patient-supplier's tissues. Another measure requires separate consent forms for the treatment of a patient's disease and for the granting of research rights over a patient's tissues. Of course, an additional form would be needed for the supplying of tissues for commercial vending whenever the likelihood of commercial value in rare substances arose. This system would also require a provision specifying appropriate sanctions for the failure to keep these different relationships separate.

Nevertheless, even the above discussion falls short of addressing the right to transfer rare tissues by sale. The deeper issue may be clarified by addressing the following question: If we can clearly separate the doctor/patient, researcher/

subject, and vendor/supplier relationships, should sale transfers of rare, replenishable fluids, such as John Moore's, be permitted? If the number of possible sales of such fluids is small (and there appear to be few rare and valuable fluids such as Moore's), then perhaps occasional sales would not pose any serious threat. However, it is quite another matter to have a thriving, full-fledged market in replenishable tissues; in fact, the possibility of such a market is what poses the greatest public concern. How will markets in human tissues affect our views of ourselves and our fellow human beings? What are the advantages and disadvantages of permitting people to sell their fluids versus the advantages and disadvantages of state interference with "people's dispositions of their physical selves"?[9]

Several considerations of normative ethics bear on these issues, but in order to clarify these points, the right to transfer tissues by sale should be distinguished from the right to receive income from tissues that are sold. These two rights are often conjoined, but there is no necessary connection between them. For example, one could permit agents to sell one's rare bodily fluids while forbidding them to retain the whole income from the sale, for it is often the right to profit from the sale of commodities that generates considerable moral dispute.

A common consideration advanced by economists and their philosophical ancestors—the utilitarians—is that incentives that allow people to sell their goods and to keep the income from this sale must be created in order for these people to produce goods that will benefit the entire community. These rights would be granted to individuals in expectation of future benefits to the community. Of course, any disadvantages to the community that might result from these rights would be considerations against granting the rights. In contrast, nonutilitarian theories often rest on personal desert as a basis of the right to sell and the right to receive income. Again, even if there are reasons of desert for granting such rights, these must be weighed against reasons that oppose the granting of such rights.[10]

To illustrate this point, consider certain utilitarian and nonutilitarian arguments with respect to natural talents, such as Mozart's ability to compose music. Utilitarians would argue that Mozart should be entitled to sell his compositions and profit from their income, since these incentives are needed if he is to develop his talents for the benefit of the community. On the other hand, nonutilitarian, desert theorists would argue that Mozart is entitled to sell the products of his talents and profit from their income because he has exerted the effort to develop his talents. However, both arguments lose their force when applied to the context of natural body fluids. (I ignore arguments of self-interest.)

The extent to which incentives are important in the present context depends upon whether any benefits will be provided if the rights to sell and to receive the income from sale are not permitted. Incentives are important if we wish to encourage efforts to produce the commodities, but in the case of providing or not providing valuable body tissues, little effort would be required. There would be little need for incentives, except to induce people to provide blood withdrawals in cases such as Moore's, and few incentives seem necessary for this. Such procedures are not terribly risky, intrusive, or painful, can be accomplished easily on an outpatient basis, and can be achieved relatively quickly without much disruption of one's life.[11]

A worst-case scenario might be one in which an individual has body fluids that are extremely valuable to many people, and yet he or she remains unwilling to donate them, except for a high sale price. In this case, both the right to sell and the right to receive income might be needed as incentives. Different solutions would be needed for each particular case, but at present these incentives seem unnecessary. Even if reasons for such incentives did exist, they would have to be weighed against any potentially negative consequences that might result from the granting of such rights. For example, if people like John Moore can sell their body tissues, how will this affect our view of ourselves and our institutions?[12]

The arguments based on desert fare no better. In the case in which fluids are naturally part of one's body, one can neither claim the right to sell these fluids nor the right to receive the income from them by asserting that one has, through effort (a typical desert basis), earned these rights. No, or hardly any, effort is needed to provide such fluids. Yet, a different kind of desert theorist might argue that there *is* some natural desert, such that the owner of a valuable thing ought to be entitled to sell it and to benefit from its income, even if he or she had exerted no effort to develop the object, or to make it more valuable. In short, mere ownership is enough to merit the rights to sell and receive income from selling the object—that is, those who benefit in the natural lottery should be entitled to benefit from their own good fortune.

This view raises one final and important point. In his *Theory of Justice*, John Rawls argues that in social distributions of wealth, greater amounts should attach to advantages that people possess only when it is reasonable to expect that greater rewards will motivate advantaged individuals to exert efforts that will benefit everyone, especially those who are most in need.[13] In short, we might use Rawls' arguments to say that when people have benefited in the natural lottery, for example, by having rare or unique body fluids, they should be rewarded only if it can be expected that such rewards will benefit the community as a whole. Yet in actuality, this is an incentive argument in disguise, and as shown earlier, such arguments rest on dubious premises in cases where individuals have rare body fluids. Even if there were a weak desert argument for granting individuals sale and income rights, this would have to be weighed against any negative consequences that might result from such rights. Thus, it seems difficult indeed to justify such rights, given the present evidence.

Before affirming this conclusion, however, consider some issues of distributive injustice. Distributive injustice might result when doctors, rather than their patients, profit from the use of the patients' tissues. In John Moore's case, it is clear that Moore had been treated unjustly. But did this injustice result because he did not consent to the commercial production of his fluids, or because he was denied an income from the sale of his tissues? The former seems more likely in this case. We can still question the basis of Moore's distributive claim. Both desert and incentive reasons seem implausible. I do not claim to have exhausted the arguments in favor of having rights to the sale of and income from human tissues, but I have sketched the most plausible of them.

Although the arguments I have outlined do not support a right to sell or a right to receive the income from the sale of body tissues, in cases in which people need incentives to contribute valuable body fluids for the benefit of the community, we must be imaginative in designing payments to elicit such contributions. For example, following Rawls we might permit an individual to sell her rare body fluids to bioengineering firms without permitting her the right to the entire income from such a sale. Instead, one might contribute the difference between the full sale price and the amount needed to elicit her cooperation to a community research fund that would benefit the community as a whole. Thus, she would be given the right to sell her body fluids, but her right to receive income from the sale would be limited. Such an arrangement might avoid some of the worst aspects of a market in human tissues.

The ambiguity of Moore's patient/doctor relationship and that of the intended uses of his blood bear on the validity of any sale of fluids that he might agree to. It seems that even if the right to sell and the right to receive income were granted to possessors of especially valuable body fluids, the validity of such sales ought to be contingent on whether an individual has been fully informed by his or her physician and whether he or she has given voluntary and uncoerced consent. The case for requiring informed consent when the aim is to try to ensure that people do not mistakenly forgo benefits is somewhat weaker than the case for informed consent when the aim is to try to ensure that people do not mistakenly consent to harms (Murray 1987).

To the extent that ambiguities exist in doctor/patient relationships and in the intended uses of patient's tissues, such as those in Moore's case, the above requirements cannot be satisfied. Consequently, whether legislators agree to grant people the right to sell their tissues and to receive the income from this sale, a provision should be made to permit sales only if the commercial supplier/vendor relationship is clearly separated form the patient/doctor and subject/researcher relationships.

Conclusion

In the preceding sections I have sketched some of the major moral issues that should be addressed in considering the patenting and sale of an individual's body tissues. We should, nevertheless, realize the limitations of this discussion. The patenting and sale of replenishable, nonvital blood serum or whole blood is a much easier case to justify than the sale of nonreplenishable or vital fluids, tissues, or organs. Cases like John Moore's, in which an individual has unique or very rare fluids that are patentable and valuable to others, are likely to be few in number. Furthermore, to the extent that such fluids, like gold or other precious metals, are greatly modified by others, Moore's claim to benefit commercially from the sale of his plasmids is greatly weakened or might be extinguished altogether.

In general, the very special conditions of the Moore-hypothetical case make it a poor model for cases that depart from it in important respects. To the extent that I have considered recommendations about what should be done, even in Moore's special case, I have suggested that there is a weak to nonexistent prima facie case to permit people the right to transfer their tissues by sale and the right to receive income from such a sale. If it turns out that there are more uniquely valuable tissues or fluids than biologists currently estimate and that these materials can be patented (or if the federal and state laws are modified to allow patenting of nonunique tissues), and if the biotechnological revolution forces more cases like John Moore's upon us, then such a prima facie case may be strengthened. However, even if there is a prima facie case for granting these rights, we should be sure that the negative consequences that presently challenge the granting of such rights are indeed outweighed.

Notes

1. Moore's testimony before the Investigations and Oversight Subcommittee of the U.S. House of Representatives Committee on Science and Technology, Hearings on the Use of Human Patient Materials in the Development of Commercial Biomedical Products, 99th Cong., 1st sess., 29 October 1985.

2. Ibid., quoted from the consent form that Moore made available at the congressional hearings.

3. Personal communication with one of Moore's attorneys, from the firm of Gage and Mazursky of Los Angeles, 29 October 1985.

4. Moore's testimony before the Investigations and Oversight Subcommittee.

5. In his oral testimony, Moore did not explicitly raise the issue of payment for his blood plasma, only the issue that he was not informed about uses to which his withdrawn blood was put. However, in the reported counts of his civil suit are the equivalents of claims to property rights in his body tissues. One count of his complaint is that Dr. Golde and UCLA "converted" his property for their own purposes. The civil wrong of "conversion" is difficult to define accurately, but one account characterizes it as "an act of willful interference with a chattel, done without lawful justification, by which any person entitled thereto is deprived of use and possession" (Prosser and Keeton 1984).

6. Telephone communication with H. Moatz, patent examiner, 5 November 1985. Despite the uniqueness criterion, however, even if interferon, one of Moore's blood products alleged to have been patented by UCLA, is in *purified* form, it may be patented; it is novel, for it does not naturally occur in nature in purified form.

7. Professor S. Munzer of the UCLA Law School in unpublished writings has called my attention to the difference between these questions. One might have rights in one's body, call them "personal rights," but we would be reluctant to say that one had or should have property rights in one's body.

8. The rights listed correspond to a list given by Becker (1977, p. 19).

> (1) to exclusive physical control of the thing owned; (2) to personal enjoyment and use [distinct from the right to manage and the right to the income from *X*]; (3) to decide how and by whom a thing shall be used; (4) to the benefits derived from the foregoing personal use of a thing and allowing others to use it; (5) to alienate the thing and to consume, waste, modify or destroy it; (6) immunity from others' expropriating it; (7) to devise or bequeath the thing; (8) to indeterminate length of one's ownership rights; (9) a duty to forebear from using the thing in certain ways harmful to others; (10) liability to having the thing taken away for repayment of a debt; and (11) [conformity to existing] rules governing the reversion of lapsed ownership rights.

9. See Bermant, Brown, and Dworkin (1975), for a summary of some positions on this issue.

10. With regard both to utilitarian and nonutilitarian positions, I am relying on a standard philosophical distinction between a *reason* for a certain course of action (which establishes a prima facie case for a policy) and what a person *ought* to do, all reasons or considerations taken into account.

11. Throughout I have focused on the sale of replenishable, nonvital substances, such as Moore's blood plasma. Of course, most of my comments would have to be changed, including the discussion of these past few paragraphs, were these assumptions modified.

12. Consider some of the negative consequences suggested both by Murray (1987) and by Bermant, Brown, and Dworkin (1975).

13. This is a very general formulation of Rawls' "difference" principle; for more precise versions, see Rawls (1971).

References

Becker, L. 1977. *Property Rights: Philosophic Foundations*. London: Routledge and Kegan Paul, p. 19.

Bermant, G., P. Brown, and G. Dworkin. 1975. " Of Morals, Markets and Medicine." *Hastings Center Report* 5 (February): 14–16.

Hohfeld, W. N. 1923. *Fundamental Legal Conceptions*. New Haven: Yale University Press.

Los Angeles Times. 1988. "Patient's Blood, Tissue Belong to Him, Court Says." *Los Angeles Times*, 22 July, pt. 2, p. 1.

Munzer, S. 1985. "Body Rights and Property Rights." Paper given at International Conference on Social Philosophy, Colorado College, Colorado Springs, 9 August.

Murray, T. 1987. "Gifts of the Body and the Needs of Strangers." *Hastings Center Report* 17 (April): 30–38.

Prosser, W. L., and W. P. Keeton. 1984. *Prosser and Keeton on Torts*. St. Paul, Minn.: West Publishing, p. 88, n. 2.

Rawls, J. 1971. *A Theory of Justice*. Cambridge: Harvard University Press, pp. 75–83, 101ff., 179, 316–321, 511, 585.

Critiques

PART IV

The final section of this volume highlights the broad themes underlying the case studies. The authors scrutinize the notion of "owning ideas" and emphasize how misleading it is to speak of formal ownership of any but a very limited range of physical representations or applications of ideas.

Two of the papers pursue the suggestion that intellectual property policy can be justified on the basis of the innovator's natural rights to his ideas. They show that this view must be virtually rejected. If there are no natural rights to intellectual property protection, one looks to social consequences to justify such protection. There are, however, also formidable obstacles to result-oriented justifications. The chapters in this section reassess the established philosophical and legal basis for our intellectual property system and focus upon how the system must respond to such changes as increased ties between universities and commercial enterprises. The final essay offers an original conceptual framework broad enough to encompass all the candidates for proprietary protection and the mechanisms so far devised for dealing with intellectual property. This analysis yields a structure for further reading and discussion of the normative issues of intellectual property protection. An annotated bibliography of the literature follows.

Arthur Kuflik's philosophical analysis focuses on two main questions: Does one have a natural right of ownership to the ideas one is first to conceive? And how is intellectual property policy shaped by the need to balance three basic values—free thought and expression, creativity, and fairness. He tests the hypothesis that we have a natural right to our creative ideas against specific rules and features of our intellectual property system. Since the system's rules so clearly emphasize our independent rights to think about or discuss the insights embodied in inventions, they suggest that there are no basic property rights to ideas. At most, the rules accommodate limited rights granted in order to further social welfare.

Kuflik considers several candidates for a natural-rights foundation for an intellectual property system. These include 1) a right to freedom, 2) a right to the products of one's own mind, 3) a right to privacy combined with a right to make contractual agreements, 4) a right to just deserts for effort or acccomplishment, and 5) a right to fair treatment.

Appeals to personal freedom ignore the central feature of property arrangements: they grant owners authority or control over potential users. A supposed natural right to the products of one's own mind cannot account for the exclusionary aspects of our intellectual property system. Indeed, that supposed right would entail that others who arrive at the same idea on their own are equally entitled to it. The appeal to a right of privacy invokes the right of an inventor not to disclose his invention. This right the inventor exchanges (in a "contractual agreement" with society) for exclusive control over the production and distribution of the invention. This exchange, however, cannot justify the relatively strong proprietary protection conferred by a patent.

Kuflik shows that patents also cannot be justified as just deserts for efforts or accomplishment. Our system neither rewards unsuccessful effort nor independently successful efforts. He raises serious questions as to whether desert should play any important role in the design of legal and political arrangements. His final analysis looks at appeals to fairness. In the absence of proprietary protection, others can exploit an innovator's ideas without incurring the innovator's development and research costs. This free riding is objectionable because it discourages innovation and because it is simply unfair. But the unfairness is not a matter of a right to "exclusive ownership of ideas" but of appropriate distribution of benefits and burdens. Kuflik stresses that fairness does not dictate a system giving the first inventor temporary monopoly power, and this system is itself unfair to independent researchers. Tinkering with the system to correct for this "unfairness" would be very difficult and expensive.

Michael Davis continues the discussion of a natural right to property. He agrees with Kuflik that the traditional conception of such a right comes into conflict with patent law insofar as patent law assigns exclusive rights to whoever happens to arrive at the idea first. He suggests an alternative theory of a natural right to property that is based on the Constitution and is consistent with patent law. This theory's natural property right is a much weaker claim against the government than that of the traditional conception of a right to property in "the state of nature."

Davis bases his discussion on the Fifth Amendment, which forbids Congress to take property "for public use without just compensation." He argues that our natural property right requires only that the government should not deprive us of something to which we have already been granted property rights. This limited "natural" right is in line with the Constitution's rationale for intellectual property protection, which is result oriented and emphasizes incentives for innovation. Such a limited right is consistent with other evidence of the views of the framers of the Constitution: they never include, among natural rights, a right to property in the sense of a moral right to something in "the state of nature."

In a very brief critique, Gerald Dworkin takes up the suggestion made by Kuflik at the end of his paper that the justification for intellectual property protection must be sought in its results. Kuflik raised questions about how well our system serves the purpose of bringing forth innovation in different areas of technological development and different industries. He also underlined concern about the extent to which intellectual property protection interferes with healthy competition. Dworkin amplifies some of these doubts. He emphasizes the lack of empirical evidence to support the claims made for the intellectual property system and the difficulty of getting appropriate evidence. It is notoriously difficult to test counterfactual claims such as those about how we *would* be affected by modifications in or abandonment of intellectual property controls. He insists, however, that we must remain skeptical when presented with unsupported claims about what the system accomplishes.

Dworkin points out that the system is expected to achieve a large number and variety of goals. Along with technological innovation, the list of ends includes more money for universities, quality control for new products (especially drugs), maximization of wealth, maintenance of openness within the scientific community, and preservation of the values of the university. This multiplicity of ends introduces enormous complexity. The challenge to testing becomes greater because of the need to test counterfactual claims about the effects of intellectual property protection in a system of multiple ends. Moreover, it requires that we deal with the problem of making trade-offs among the ends. In view of likely disagreements, we face the task of determining appropriate institutional settings for reaching authoritative decisions about which ends to pursue. That decision, Dworkin reminds us, must be responsive to our evidence about how to promote those ends most efficiently and justly.

Leonard Boonin reviews the philosophical and legal basis for recognizing property rights in knowledge. In the light of that review, he examines the tensions that have arisen within universities between the traditional values of scientific research and the legal norms associated with intellectual property. In his look at the legal basis, Boonin observes that the patent system fails to provide much incentive for basic research, whether for reasons of principle or because of the difficulties of administering a system that rewards theoretical research. It would be a formidable task to decide what kinds of theoretical knowledge are patentable, to resolve conflicts over priority, to determine instances of infringement, and the like. He proposes that a tax be put on the profits from patented inventions to establish a fund to support basic research.

Since basic research is not protectable, most commercial enterprises do not engage in it and it remains the province of academic institutions. Universities have nonetheless strengthened ties to corporations, in part because they have come under pressure to find alternatives to government support for basic research. Joint ventures, especially in biotechnology, have proliferated with great rapidity, drawing academics into fields promising economic rewards. This has given rise to a debate over possible commercial influence on the direction of research, threats to established practices in basic research, and erosion of academic ideals. Boonin expresses concern, for example, over the action of some states in broadening their concept of trade secrecy to encompass basic research within the university. Although he recognizes that it is difficult to evaluate these new options for intellectual property protection, he underscores a need for both parties to these new relationships to develop policies to protect institutional integrity. Boonin urges that the basic values of justice, fairness, and social utility should govern these policies, and that the (perhaps distinct) values of community be reflected in university policy.

The final essay by Patrick Croskery, which precedes his bibliography, offers a conceptual guide to the literature. Stepping back from the debates examined in previous papers, Croskery proposes a comprehensive structure for comparing, in ethical terms, the various accounts of intellectual property. He begins by assuming that there is a class of goods (in the economist's sense of *goods*) captured by the term *intellectual property*. He finds two aspects of so-called IP goods especially pertinent. One depends upon the distinction to be made between "rival" and "nonrival" consumption. The distinction depends on whether one person's use does or does not diminish or preclude the use of the same good by another. Alternatively, we may ask whether the good—a text, for example—is or is not consumed in use. A second aspect derives from the distinction between created and discovered goods. These two distinctions generate four possible characterizations of IP goods, each paired with a specific mechanism for handling the goods. For example, the mechanism of market exchange handles rival, created goods, such as cars or matches. Croskery then shows how the classification of goods and mechanisms applies to the specific ethical issues considered in the papers preceding and those cited in the bibliography.

To see how Croskery's scheme can be used, consider the category of rival, discovered goods, which includes such natural resources as coal. For assigning rights to these goods, a "prospect" mechanism is generally used. This system assigns rights to the one who demonstrates, according to some formal procedure, the prospect of finding the goods. A legal scholar, Edmund Kitch, has made the influential proposal that the patent system can and should be viewed as a prospect system. Croskery points out that when IP goods are thought of as rival, discovered goods, property rights to IP goods are justified in the same way as those to such resources as coal.

Croskery suggests that the university research system, with its distinctive pattern of incentives, handles nonrival discovered goods, such as laws of nature. Though this system is not intended to handle IP goods, it frequently turns out that it becomes enmeshed with them. Croskery's scheme thus provides an illuminating perspective from which to examine the increasing entanglement of the university research system with IP goods as university-industry research relationships multiply.

The bibliography reflects a comprehensive survey of the literature in all relevant disciplines and a careful selection of items that contribute to a normative understanding of intellectual property. Implications of intellectual property protection for research and development in science and technology receive emphasis.

11. Moral Foundations of Intellectual Property Rights

Arthur Kuflik

Patents and copyrights are among the most conspicuous examples of what is authoritatively classified as *intellectual property*. With equal authority, however, it is also said that nobody can legitimately patent or copyright an *idea*.

There is something of a puzzle here. For if ideas cannot be patented or copyrighted, then in what sense do patents and copyrights secure or protect intellectual property? A moment's reflection on this puzzle only leads to other, morally more significant, perplexities: Would the practice of granting a person proprietary rights to an idea be morally defensible? If intellectual property law does *not* make a person the owner of an idea, then to what do patentees and copyright holders have proprietary claim? And on what basis?

If one listens to what some of the staunchest defenders of private property have had to say about intellectual property, the puzzlement is likely to be exacerbated, not alleviated. On the one side, one might hear that "patents are at the heart and core of property rights . . . once they are destroyed, the destruction of all other rights will follow automatically, as a brief postscript" (Rand 1967). On the other side, one might be told, "Patents . . . invade rather than defend property rights" (Rothbard 1977).

In what follows, I address two issues: First, do patents and copyrights create (or secure) property in ideas? And second, is the practice of assigning patents, copyrights, and other forms of intellectual property morally defensible? And I argue for two theses: First, the intellectual property system cannot be satisfactorily grounded in the principle that a person literally owns, as a matter of natural right, the ideas that he is the first to conceive. And second, underlying, and to some extent shaping, the practice of granting patents, copyrights, and other forms of intellectual property is the need to strike a suitable balance among three important considerations: freedom of thought and expression, incentive to authorship and to technological innovation, and fairness.

Intellectual Property Law and the Ownership of Ideas

What Is Owned, If Not Ideas?

Do patents and copyrights bestow ownership of ideas? And if they do not, to what do they give their holders title? Federal law makes it perfectly clear that what is copyrighted is not an idea, but the particular expression that it has been given. Thus, United States Code 17, section 102 reads:

> (a) Copyright protection subsists . . . in original works of authorship fixed in any tangible medium of expression . . . (b) In no case does copyright protection for an original work of authorship extend to any idea, procedure, process, system, method of operation, concept, principle, or discovery, regardless of the form in which it is described, explained, illustrated, or embodied in such work.

But what about patents? Do they secure property in ideas? To secure a patent one must be able to specify a new, useful, and nonobvious process, machine, manufacture, or composition of matter and to do so in such detail as would enable any person skilled in the relevant "art" or discipline "to make and use the same" (35 U.S.C. secs. 102, 103, 112).

Here the term *process* refers to a method for transforming or reducing a physical substance to a different state or thing; it does not refer either to a method of thinking or of solving intellectual problems or to a method of doing business. Indeed, abstract ideas, mental processes, methods of thinking or of solving intellectual problems—no matter how new and original they might be—are not proper subject matter for a patent application (Gottschalk v. Benson, 409 U.S. 63 [1972]).

In light of all this, it is tempting to suggest that what a person patents, and thereby comes to own, is not simply an idea, but a useful or practical idea. But this theory does not quite fit the phenomenon it is intended to explain. There are two objections to it. First, having a useful idea—even granted that it is not only new but also nonobvious—is not a sufficient basis for holding a patent. Second, patenting, even when one has a sufficient basis for it, does not literally give one ownership of an idea.

Being the first to put forward a useful idea is not a sufficient basis for holding a patent. Consider the following dialogue:

> "I've just come up with a brilliant idea: I've noticed that snow melts at different rates on different kinds of surfaces. Now, imagine a substance you could spread over the sidewalks so that whenever it snows, the snow melts almost as soon as it falls!"
>
> "What is that substance?"
>
> "I don't know, but as the first person to think up this very clever idea, I'm going to patent it; then I can draw royalties from anybody who does manage to find a substance that does the job I have in mind."

Clearly, if the useful idea—brilliant and original though it may be—concerns the general function or purpose that some (as yet unspecified) device, substance, or process would serve, it does not provide a sufficient basis for holding a patent.

Granted that a person cannot get a patent merely by virtue of being the first to conceive a useful function, one might suppose that contributing new, nonobvious, and useful ideas about how the specified function is to be performed would qualify someone for a patent.

But then consider the following—someone discovers the special theory of relativity. Pondering $E = mc^2$, he realizes that it may be possible to derive significant amounts of energy from matter. He suggests that the heaviest, most unstable elements—uranium, for example—are likely to provide the most promising material basis for effecting such a conversion. Though he has practical insights indispensable to the development of an extremely important technology—insights for which others might be more than willing to pay a handsome price—this person does not have a sufficient basis for a patent.

Persons who put forward new and nonobvious ideas indispensable to the development of new and useful technologies are not rewarded by the patent system. Only those who go further and offer specific instructions about how to compound a useful chemical substance, engage in a productive process of manufacture, and so forth are entitled to the prerogatives of a patent holder. Moreover, these instructions must be sufficiently clear and precise to enable persons skilled in the relevant art or discipline to replicate, without further experimentation or invention, what has been specified.

Even when one has a sufficient basis for a patent, it does not literally give one ownership of an idea. Imagine that someone has not only envisioned a function to be performed, but has also conceived, and in detail sufficient to enable others in the field to "make and use" the same, something that is capable of performing that function. And suppose he has obtained a patent. The fact remains that anybody has the right to think the thoughts that characterize whichever design, formula, or process he has conceived. Thus, anybody has the right to believe that if certain materials are put together in a certain way one will have something (whether it be a machine, or a manufactured product, or a chemical compound, or what have you) that is capable of performing the designated function. Nobody needs the permission of the inventor either to hold such beliefs or to discuss them with others. Thus someone who can specify a new, useful, and nonobvious machine, process of manufacture, or formula can obtain the right to exclude others from making, using, or selling anything that meets that specification. But he cannot prevent them from thinking about, discussing, and otherwise deriving inspiration from the practical insights that underlie his invention.

To sum up, what qualifies a person for a patent is not that he has an idea—even a useful idea—but that he has a useful idea of a highly specific and practicable sort. That is, it is the design for a machine or mechanism, the formula for a composition of matter, or the process for the transformation and reduction of a physical substance to a different state or thing. And what he comes to own, or indeed monopolize, is not the idea as such but, for a limited period of time, the right to "make, use, or sell" that which answers to it.

Freedom of Thought and Speech as a Constraint on Intellectual Property Rights

There is a parallel here between copyright and patent. Just as the person who holds a copyright does not have a proprietary right to an idea, but to a particular tangible expression of it, so it might be said that the patent holder does not have proprietary claim to the useful ideas behind his invention, but rather, to their actual practical application.

It would be a mistake to suppose that this observation holds only idle intellectual interest. For underlying the fact that ideas as such can be neither patented nor copyrighted is a fundamental moral concern: the rules of the intellectual property system must not be formulated in ways that might jeopardize freedom of thought and speech.

Other important features of intellectual property law attest to this same concern. Thus, patentability does not extend to scientific laws or to methods for solving mathematical problems. As the Supreme Court has ruled, these are the "basic tools" of scientific and technological research and cannot be preempted by anybody (Gottschalk v. Benson [1972]). Also relevant to the present point is the fact that the specification of a granted invention must be placed in the public record, in "full, clear, concise and exact terms" (35 U.S.C. sec. 112). In virtue of this, others have the opportunity to assimilate and draw inspiration from the inventor's insights.

Turning to the laws governing copyright, one finds that the rights of the copyright holder are delimited by the "fair use" doctrine under which a work may be reproduced "for such purposes as criticism, comment, news reporting, teaching (including multiple copies for classroom use), scholarship, or research" without infringing the copyright holder's proprietary rights (17 U.S.C. sec. 107). Nor is it an infringement of copyright "for a library or archives, or any of its employees acting within the scope of their employment, to reproduce no more than one copy or phonorecord" provided that (1) it is done "without any purpose of direct or indirect commercial advantage"; (2) the collections of the library or archive are open to the general public or to the body of scholars in the relevant field; and (3) a notice of copyright is included (17 U.S.C. sec. 108).

To make sense of such provisions and qualifications it is plausible to suggest that the intellectual property system has been so designed that, whatever the purpose to be served by granting authors and inventors copyrights and patents, the basic freedom to think about and to discuss the ideas and insights that underlie their writings and inventions needs to be protected.

Justifications for Intellectual Property Rights

As has been shown, the laws of patent and of copyright are generally formulated within a framework that is intended to preserve basic freedom of thought and expression. But why should intellectual property rights be assigned and protected in the first place? In what follows, I will first consider the question of whether the practice of granting patent rights is morally defensible, and if so, on what ground. Then, after noting an important contrast between the way in which the laws of copyright and of patent deal with the question of independently arrived at but significantly similar achievements, I will explore the question of whether the considerations that seem to provide the most significant support for the patent system support the copyright system as well.

A Libertarian Argument

One may begin by recalling the somewhat vague but provisionally appealing principle that people should be free to do as they choose so long as they do not interfere in other people's lives. Could the inventor's right to patent his invention be a simple exercise of this right to freedom? Whatever the merit of the principle, it is simply too weak to yield the desired conclusion.

Thus, consider the following: Someone invents the wheel and starts wheeling things around. Others get the idea and, after duly acknowledging and praising the person who is the source of their inspiration, make wheels of their own for their own personal use. To be sure, when the inventor makes wheels and starts wheeling things around, he does not interfere in the lives of others or limit their liberty in any way that could provide legitimate ground for complaint. But the same could be said of the others: when they make wheels for their own personal use, they are not interfering in his life or limiting his liberty to make and to use wheels.

It is tempting to object that their use of the idea does constitute an interference in his life. After all, they took the idea from him without his permission. But this objection is subject to the following line of criticism. When someone takes my car without my permission and drives it around, then all the while he is driving around, he deprives me of the personal use of it. But when someone takes my idea and—after acknowledging me as the source of his inspiration—makes use of it in his personal life, he does not thereby deprive me of the liberty to do the same, that is, to make use of the idea in my own personal life.

Indeed, there are at least three senses in which a person who gets an idea from me need not be taking it away from me: (1) I can still think it; (2) I can still enjoy whatever praise or admiration others might be disposed to give to me as the person who thought of it first; and (3) I can still use it, to all the same personal advantage, in my own personal life. Here it may be objected that if others are at liberty to use the idea without his permission, then the person who came up with the idea first will not make so much money as he would have made otherwise. So in putting it to one's personal use, one does take something away from the other person. One deprives him of something that is rightfully his.

But note that "so much money as he would have made otherwise" here signifies so much money as he would have made if he had had the authority to decide who shall use the idea and on what terms—in short, if he had enjoyed monopoly control.

Thus, to decide whether the use that other people make of an idea has deprived the person who first thought of it of something that is rightfully his, one has to decide whether the first to think of it is entitled to exclude anyone else from using the idea without his permission. Such an entitlement is not a mere liberty, but a power or prerogative: to have it is to have a measure of authority or control over the lives of others. It may be a perfectly legitimate authority, but appealing to personal freedom is not going to be sufficient to legitimatize it. One must appeal to other (presumably stronger) considerations.

The Appeal to a Natural, Inherent Property Right in the Products of One's Own Mind

Consider then the suggestion that the right to patent is not simply a matter of freedom, but an implication of the principle that a person owns the products of his own mind. On at least one reading, this principle is certainly very appealing. After all, an idea that is yours (that is, that you have thought up on your own) ought to be yours; you should have the right to think it and to put it to any use that does not violate anybody else's rights. (This last qualification applies to rights in general: my right to my knife does not give me the right to put it in your chest.) But those who argue for patent rights need a stronger argument to help them establish a stronger conclusion. They need to argue that a person not only owns (nonexclusively) the application of any useful idea that is the product of his own mind but also has, if he is the first to think up the idea and reduce it to practice, the right to exclude others from using it.

To establish this conclusion one might reason along the following lines: In giving a person exclusive right to the application of an idea that originated with him, no one else's position is worsened. Since the invention would not exist if not for him, it is and ought to be entirely his.

Perhaps the first thing to note is that if the patent system is really to be based on the principle that a person has a natural right to monopolize the application of a useful idea that he is the first to conceive, then it ought to be possible to obtain exclusive right to the application of more general ideas—for example, the idea of using electricity to provide indoor illumination, or the idea of converting unstable elements such as uranium into nuclear energy. As I have already noted, however, there are many important ideas of great practical significance whose application is not, at least under the present system, made the exlusive right of their first discoverers.

This observation leads to another, more damaging, one: If the right to patent is grounded in the principle that there is a natural right to the exclusive use of the original products of one's own mind, then there seems to be no reason that that right should not also extend (*a*) to theoretical as well as to practical ideas, and (*b*) to their public discussion as well as to their technological application. In short, the putative right, and the proposed line of argument based on it, are difficult to reconcile with freedom of thought and expression. What is needed is a coherent account of why, even though people have such a right, it applies only to certain products of their mental activity—specific inventions, particular works of authorship—rather than to all such mental products. But even if such an account could be constructed, the approach in question would still be highly questionable on at least two other counts.

First, it is implausible to suppose that someone who is the first to think up a useful idea has conceived something that would not have come into existence otherwise. Brilliant though it was, the idea of the wheel would have independently occurred to others. Proof of this is provided by the fact that the idea of the wheel did occur, at different times and in different places, to peoples who had no contact, whether direct or indirect, with one another. And of course, the same can be said, with better documentation, about more recent technological advances. But the patent system gives the first discoverer a right to exclude—for the duration of the patent term—even those who, operating independently, make the same discovery shortly afterward. Presumably, these independent inventors are equally entitled to the products of their own minds. Thus, the putative right to appropriate the product of one's own mind does not support, but actually tells against, the policy of giving exclusive rights to first inventors.

Of course, it is not always entirely clear just when a technological development would have occurred in the absence of its actual first discoverer. This might suggest something like the following line of argument: The policy of granting a seventeen-year patent term is an—admittedly often inaccurate—approximation to the period of time it would have taken others to come up with the invention on their own. Letting the patent pass into the public domain after that period of time is a way of recognizing the fact that sooner or later the continued enjoyment of exclusive rights would indeed constitute a wrongful worsening of the situation of at least some (not necessarily identifiable) individuals (compare Nozick 1974).

But if this reasoning were indeed appropriate, then it would hardly justify anything like the present system. This is because nearly contemporaneous, independent inventors could not be rightly excluded even for seventeen years. Furthermore, in cases in which the public disclosure of an invention occurs soon enough to put an end to further independent research, the policy of assigning the very same fixed term of exclusive rights, without regard to the particular invention or the general field in which it occurred, would be unjustly crude. Different areas of research and development will exhibit demonstrably different rates of overall progress. Even within a given field, progress on a particular technical problem will vary according to the stage of the field's development and the intensity of effort devoted to the problem. The principle that people have exclusive right to the product of their own mental activity, just so long as others are not made worse off than they would have been in the absence of that mental activity, would call upon society to make a scrupulous effort to obtain the best available evidence on such matters and to set up the rules of the patent system in a way that more adequately reflects these variations.

Second, whoever is the first to think up some important idea, whether practical or theoretical, he is almost certainly not drawing upon his own mental resources only. According to ancient legend, Athena sprang full-grown from the head of Zeus. But human beings do not spring full-grown from either a human or a divine parent. Certainly, they add to and enrich the life of the community in which they live, but their capacity to do so, as well as the more particular ways in which they do it, are made possible by a shared and historically transmitted hertiage of language, culture, experience, and craft. When hailed as a great and original genius, Isaac Newton responded that he was, after all, only standing "on the shoulders of giants." Indeed, even in making this admirably humble remark, Newton was standing on the shoulders of others; the phrase was not original with him but had a long and illustrious history of its own (Merton 1967).

Thus, from a putative right to the products of one's own mental activity it does not follow that anybody can rightly claim exclusive control over a useful invention that he is the first to conceive. For nobody can rightly claim that a useful invention, or indeed any intellectual achievement, is fully and solely the product of his own mental activity.

Right to Privacy and Freedom of Contract as the Basis for Patent Rights

As I have already noted, an important feature of the patent system is that the applicant must make a disclosure of his innovation in such detail as would be sufficient to enable "any person skilled in the relevant art to make and use the same." This may suggest something like the following line of argument.

The right to privacy implies that an inventor has the right not to disclose his invention. Patent right—the right to exclusive control over the production and distribution of the invention—arises as part of a contractual agreement between the inventor and the government. The inventor discloses his invention in return for being granted a (limited) monopoly privilege. In virtue of this bargain between society and the inventor, the inventor comes to have the right to exclude others from making, using, and selling the invention in question. On this view, patent rights are not basic rights but they are the legitimate product of the exercise of two other rights: the right to privacy (which implies the right not to disclose any details about one's invention) and the right to make contracts.

A crucial objection to this line of argument begins with the observation that freedom of contract is not unlimited: a person has no right to make a "hit" contract for example. Thus, to decide whether the would-be patent holder can legitimately demand that nobody else—not even near-contemporaneous independent inventors—be allowed to make, use, or sell whatever is in question, one needs to know if he has the right to make such a demand. If what is demanded is illegitimate, then freedom of contract will not somehow bestow legitimacy upon the corresponding concession. Thus, to show that monopoly privilege is a legitimate demand, one cannot merely appeal to the right to privacy and the right to make a contract. The relatively strong proprietary right involved in holding a patent can only be justified by appeal to some other, presumably stronger, consideration.

Patent Right as a Matter of Just Desert

In order to provide the added justificatory strength, it is tempting to invoke the notion of just desert. On this approach, the power or prerogative that is afforded by a patent is legitimate insofar as it is deserved. Deserved in virtue of what? Possible candidates are effort and accomplishment. In either version, the principle that people ought to be rewarded according to what they deserve would prescribe more than it seems reasonable to do.

A principle of desert for effort would imply that unsuccessful researchers who nevertheless have expended a great deal of effort and money in an earnest attempt to come up with something useful to the public, and are therefore very deserving, ought to be rewarded. But the patent system does nothing of the kind. Nor does it seem plausible to suppose that it should. A principle of desert for successful accomplishment would imply that independently successful inventors also ought to be rewarded.

Whatever the basis for desert, there is the further problem of fixing the *size* of the deserved reward. How much of a reward does an innovator deserve (whether for his effort or accomplishment)? It is difficult to believe that, regardless of effort or accomplishment, the innovator's deserved reward is whatever income he can secure through holding and exercising a seventeen-year monopoly.

Finally, and more generally, it is far from clear that desert is an appropriate basis for the design of legal and political arrangements. What people deserve is often quite properly contrasted with the (institutional) entitlements that they (morally) ought to have. A baseball team may rightfully lay claim to a victory that was really deserved by the other team. Why the contrast? If both teams play fairly and in full observance of the rules, then the team that actually wins is the rightful victor. But if the other team both has the greater talent and has made the greater effort then it might be said to be more deserving of a victory. Why then did it lose? "Bad luck," one might say. Of course, who is to decide which team is more talented or has made the greater effort?

An institutional arrangement that superimposed upon its system of announced rules and regulations an authority with the discretion to determine who is really most deserving after all, and to award victory accordingly, would not seem morally defensible. The discretion in question would be too susceptible to arbitrary or discriminatory exercise. It is not that the notion of desert has no meaning. Rather, if one is to think of it as a principle of institutional design, it seems more appropriate for God or some other supposedly incorruptible and omniscient being than for ordinary mortals.

From these reflections, this point emerges: no plausible conception of what people deserve and why they deserve it would lead to anything like the present patent system. It is, in any event, questionable whether the notion of desert ought to play a significant role in the design of legal and political arrangements.

Progress in Technology: A Forward-Looking Defense of Patent Rights

Perhaps the most plausible argument for the special authority that is vested in patent holders turns on the long-term effects of the patent system upon research and development efforts. The suggestion is that, as an incentive to greater technological progress, the normal condition of free and open competition may need to be, from time to time and for a limited period of time, suspended.

In this spirit, the U.S. Constitution in article 1, section 8 does not call upon Congress to make laws protecting a person's natural proprietary right to the products of his own mind. Instead, as is well known, the Constitution authorizes Congress to enact laws whose purpose is "to promote the progress of science and useful arts, by securing for limited times to authors and inventors the exclusive right to their respective writings and discoveries." The basic philosophical point is elaborated by the Supreme Court:

> The patent monopoly was not designed to secure to the inventor his natural right in his discoveries. Rather, it was a reward, an inducement, to bring forth new knowledge. The grant of an exclusive right to an invention was the creation of society—at odds with the inherent free nature of disclosed ideas—and was not to be freely given. Only inventions and discoveries which furthered human knowledge, and were new and useful, justified the special inducement of a limited private monopoly. (Graham v. John Deere Co., 383 U.S. 1, 9 [1966])

Thus the patent system emerges as a device for getting the best of both worlds. Monopoly privilege serves as an initial incentive to innovation; its limited duration eventually allows for the usual effect of free and open competition. Moreover, all this takes place within a framework that preserves the basic freedom of thought and speech so essential to the long-term progress of both science and technology.

There is a good deal of common sense in this line of argument. Those who engage in research and development often have to expend significant amounts of time, energy, and money without much assurance of success. Moreover, those who do succeed face the prospect of being undersold by competitors who are able to discern and duplicate what is usefully innovative without having to incur comparable research and development expenses.

In virtue of these two difficulties—the greater uncertainty of success and the relative ease of free riding—research and development efforts are likely to fall short of what the long-term health and well-being of society would seem to warrant. The patent system can be viewed as a device for correcting, at least to some extent, for these difficulties. Does it correct enough, or perhaps too much? Some would claim that the patent system overstimulates technological innovation and fosters wasteful duplication of research effort. Others would argue that the incentive it provides is not strong enough.

To evaluate such complaints one needs to be able to measure the impact of the patent system upon the rate of technological development. The state of affairs that would have obtained were patent rights not actually recognized has to be evaluated against the state of affairs that does obtain in virtue of them. It is not easy to verify or validate this rather complicated counterfactual comparison. Moreover, one needs to know more about what rate of technological development is supposed to be optimal and why. It is one thing to maintain that under the normal operation of market forces, research and development efforts would surely be inadequate, yet quite another to claim that one can specify with any precision an optimum level of such effort.

Now, there may well be cases in which—without knowing just what level of research would be optimal—one can nevertheless be reasonably confident that more research than is presently being undertaken would be desirable. This hardly constitutes a fatal criticism of the practice of recognizing patent rights as such. If greater incentive to research and development is needed, it can generally be achieved through modifications of the patent system itself (for example, extending the life of the patent, granting the patentee the right to make licensing agreements that bar challenges to the legal validity of the patent) or through additional mechanisms (government research grants, prizes) that can operate in conjunction with the patent system.

Of course, in evaluating an institutional design or public policy, one must look not only at the prospective benefits but at the costs as well. Competition in the marketplace is generally regarded as a spur to higher quality of production at lower prices. Monopoly is thought to be counterproductive of these good effects. Thus, whatever contribution the patent system makes to the progress of technology needs to be weighed against the reduction of quality and the increase in price that are the usual consequences of monopoly privilege.

In addition, it seems likely that the supposed benefit of having the patent system—namely, incentive to innovation—will vary considerably along with the nature of the technology. Securing a patent tends to be a prolonged, costly, and uncertain process; once a patent has been obtained, the effort to protect it through infringement suits can also be costly and prolonged. Thus, for fields in which there is rapid technological development, patent rights may bring too little too late to provide any real incentive. In these areas, simply getting there first may be its own, and the most significant, reward.

Even so, the rate of technological innovation has certainly been greatest in those social systems which do recognize intellectual property rights. It has yet to be demonstrated that other factors—cultural rather than legal—have played the more significant role. In the absence of such a demonstration, it seems highly unlikely that, even without a measure of intellectual property protection, technological progress would have been just as great.

Moreover, the alleged conflict between providing a healthy incentive to innovation and maintaining a vigorously competitive marketplace is not so clear-cut as might appear. Once again, much depends on the particular field or industry. There are areas in which significant research and development can be meaningfully undertaken by relatively new and smaller firms. Failure to provide some measure of exclusivity to their accomplishments may only ensure that such firms have little chance of surviving, no matter how innovative they are. Without such protection, the Goliaths of the industry could readily assimilate any commercially viable innovations and bring them to market at prices that the smaller firms cannot match. In some fields, then, limited monopoly protection may not only spur innovation, but actually help the Davids to establish themselves against the Goliaths. The net result, of course, would be to widen and invigorate, rather than to weaken, the competitive field.

On the other hand, there are fields in which technological change comes mainly from very large firms that have invested heavily in research and development too costly and complicated for newer and smaller firms to handle. In these areas, there may be little chance for the field of competitors to widen—unless other firms do have the guaranteed opportunity to bring innovations to market, while paying reasonable royalties to the innovating firm. An obvious problem here is to determine a reasonable royalty rate. But if some policy of this sort could be put into practice, it might represent an appropriate balance between the need to encourage innovation and the need to keep markets in new technologies reasonably competitive.

The Appeal to Fairness

A useful invention can make a positive contribution to the good of others. To arrive at it, the inventor(s) may have to expend a considerable amount of time, energy, and money. Sometimes, other people come along and—being in a position to imitate, duplicate, or reverse engineer the invention without sustaining comparable research and development costs—produce the same, or an obviously similar, product at a lower price. By free riding on the efforts of the original discoverer, they achieve a superior competitive position. It seems unfair that the persons whose efforts have helped to make a technological benefit possible are, by very reason of those efforts, placed at a significant competitive disadvantage.

Of course, as has already been seen, free riding can be worrisome—not because it is inherently unfair, but (from a more purely forward-looking or consequentialist perspective) in virtue of how it weakens the incentive to engage in innovative research and development in the first place. An interesting question, then, is the extent to which free riding can be regarded as objectionable in its own right, quite apart from its impact upon the rate of technological development. Grant, for the sake of argument, that free riding of this sort is, in some sense, unfair; one may still well wonder what would be fair?

Fairness might seem to imply that, at the very least, the persons who have shouldered the burden of making a benefit to others possible ought to receive adequate compensation. This raises the obvious question, When is compensation adequate? Unfortunately, the obvious answer—When it is enough to cover the costs of research and development—is not without difficulties of its own. Thus, it is perfectly conceivable that the time, money, and effort actually expended were excessive and that a more efficiently managed research and development project would have yielded the same result at lower cost. Alternatively, it is possible that the benefit to others—though real—is not great enough to have warranted the heavy expense of (even the most efficiently undertaken) research and development. So from the mere fact that someone has managed to produce a technological result that is beneficial to others, it cannot be inferred that he or she ought to receive a monetary return that completely covers his original research and development costs.

It might be thought that what fairness requires is not that inventors be compensated for their efforts but rewarded in proportion to the value of the contribution those efforts have made to the well-being of others. But what is the value of a given contribution? And what would count as an appropriate reward? Providing a satisfactory account of such matters would seem to be an even harder task than working out a theory of adequate compensation.

Instead of trying to answer these questions with any precision, or even at all, a plausible route to take might be to protect the innovator against blatant free riding but then to let his financial return be determined by the forces of the marketplace. He would accept the outcome whether those forces accurately reflect the long-term value of his contribution to society and whether this original investment is recovered. Taking this route avoids the unpleasant and illiberal prospect of giving someone the power and discretion to sit in Washington and impose upon the community of innovators and upon society as a whole his own particular view of what has value.

On this theory, the intellectual property system results from an attempt to achieve a measure of fairness within the limits of a safely decentralized economy. In essence, inventors are thought to be entitled—not to compensation or reward—but rather, to a fair chance to achieve a market determined return on their investment.

Two Objections

There are at least two problems with viewing the patent system in this way. First, to qualify for patent protection a new, useful, and nonobvious technological development does not have to be the result of prolonged or expensive effort. Nor, for that matter, does it have to be the upshot of an intense flash of genius. Thus one may read that "patentability shall not be negatived by the manner in which the invention was made" (35 U.S.C. sec. 103). The crucial question is whether the development would or would not have been "obvious at the time the invention was made to a person having ordinary skill" in the pertinent discipline.

To be sure, most of the inventions that have required long and hard effort to bring forth were not particularly obvious, for what is obvious is likely to be achieved with relatively less difficulty. Moreover, much of what has been achieved without any significant effort or expense has indeed been obvious. Still, even obvious developments may be brought forth only after a good deal of fairly routine effort has been expended. And at least some new, useful, and nonobvious inventions have been the fortuitous product of virtually effortless chance discovery. But hard work is not rewarded by exclusive rights unless the product of that work constitutes a nonobvious divergence from the "prior art." So if an obvious advance is costly to initiate, yet once on the market, relatively cheap to knock off, the patent system provides no special reward to the initial developer, and offers no special legal remedy against free-riding competitors.

Moreover, the fact that an innovation was brought forth with no appreciable burden (either in time, energy, or money) to its innovator would not provide any barrier to obtaining a patent. This is despite the fact that would-be imitators would not be taking a free ride at the expense of someone else's costly efforts. Indeed, if the innovation were immediately placed in the public domain instead of being protected by patent, then everyone—the first discoverer no less than his imitative competitors—would merely be riding free on the back of good fortune.

In view of all this, it is difficult to believe that the sole rationale for the patent system as we know it is to prevent unfair free riding. To be sure, the system does go a long way toward giving expense-laden innovators a fair chance to achieve a return on their investment. But evidently another, possibly more fundamental, purpose of the system is simply to induce the discovery, disclosure, and distribution of new, useful, and nonobvious technological developments, regardless of how much or how little effort and expense was involved. Those who have a taste for theoretical simplicity might be inclined to take a further step and suggest that the concern to prevent unfair free riding carries no genuinely independent weight at all. On this view, free riding is something to be concerned about only insofar as it discourages future research and development. Of course, the appeal to simplicity is not by itself sufficient. A simpler theory can also be altogether too simplistic.

Thus, the fact that exclusive rights are sometimes awarded even when it would not be unfair to deny them hardly shows that a concern with fairness has no role to play at all. It merely suggests that, in addition to being fair, the system is designed to encourage the disclosure of useful innovations, however much good fortune, rather than hard work, had to do with their discovery. By the same token, the fact that the system is not designed to prevent people from (less expensively) imitating that which is obvious to any ordinary practitioner of the relevant art or discipline does not establish that fairness is altogether irrelevant. A more plausible explanation is that while fairness is an important value, so is freedom of competition. Society is loathe to restrict that freedom unless the person whose efforts are being (to some extent) protected has done more than bring forth that which was, in any event, obvious to virtually anybody in the field.

There is, however, a further feature of the patent system that, leaving the quest for theoretical simplicity aside, seems to offer more substantive support to the view that incentive to innovation is the only plausible rationale for the system as it is presently constituted.

Researchers who, without resorting to imitation, duplication, or reverse engineering, nevertheless manage to arrive at an otherwise patentable result within a short while of its initial (as yet unpublicized) discovery are not rightly regarded as free riders. Nevertheless, the patent system as it now stands denies these independent inventors the right to make, use, or sell the invention without the permission of the patent holder. (The exact details of the American patent system are remarkably complex. If someone is the first to conceive an invention and also the first to reduce it to practice, he is indisputably entitled to patent it. However, if the first to conceive the invention is nearly, but not absolutely, the first to reduce it to practice, the question becomes whether this so-called senior inventor began to use "due diligence" in reducing his conception to practice before or after the junior inventor first conceived the invention.)

From the standpoint of fairness, there appear to be two problems with this. First, granting a share of the privileges and prerogatives of patent holder to independent inventors would not seem to be unfair to the actual first discoverer insofar as the independent inventors' success did not result from unfair free riding. Second, not granting them at least some share of the prerogatives of patent holder would seem to be unfair to genuinely independent inventors. Shall one conclude from all this that fairness plays no role in the design of the patent system—or worse, that the system is inconsistent with the demands of fairness?

It is interesting to note that there have been cases in which the very fact that a number of researchers independently arrived at a technological result within a relatively short time span was taken to provide at least some evidence for the view that the result in question was not sufficiently nonobvious to warrant a patent. But of course, nearly simultaneous independent invention is not a perfect indicator of obviousness. Significantly nonobvious developments can be, and no doubt have been, simultaneously reached by independent researchers. The apparent unfairness of not dividing the patent prerogative among independent developers of nonobvious inventions still stands in need of at least some justification.

Here two different lines of argument may work in tandem to provide significant moral support for what might otherwise seem to be an unacceptable feature of the system. One way to justify the tolerance that the system has for this kind of unfairness is to appeal to fairness itself. Because the question of independent invention is often very difficult to settle, a system that tried to recognize the claims of independent inventors would be vulnerable to a considerable amount of fraud and deception. With mere imitators masquerading (often successfully) as independent inventors, unfair free riding could become significantly more prevalent. Thus, it might be argued that in a world in which perfect fairness cannot be achieved in any event, a policy of not recognizing claims to independent invention is least unfair.

Another route is to remember that although the concern to prevent unfair free riding plays some role in shaping the design of the patent system, it is not sufficient to justify even some of the other, much less problematical, aspects of the system. Notable among the latter is the fact that since patentability shall "not be negatived by the manner in which the invention was made," even an effortlessly arrived at lucky discovery may qualify for patent protection. Thus, if one is to make sense of, and to defend, anything like the present system, one must inevitably appeal to the way in which the system fosters technological progress.

From this perspective, the practice of exclusively rewarding whoever was the first to bring forth a nonobviously new invention is by no means unreasonable. For one thing, a system in which claims to independent invention had to be assessed and, if verified then honored, would be more costly to operate and would generate even more time-consuming litigation than the present system. In consequence, such a system would quite likely weaken rather than strengthen the morale of those who engage in research and development. In addition, and perhaps more important, a patent system that refuses to reward any but the first successful inventor is a system that provides researchers with a strong incentive to proceed as expeditiously as they can.

One should, however, take note of a possible counterincentive. Insofar as the system offers no reward for finishing second, it will discourage would-be inventors from initiating research efforts they cannot be reasonably confident of successfully completing before their competitors. (In the U.S. system sketched here, this disincentive would still be present but might not be so pronounced.) The net effect might be less research and development effort overall and a slower rate of technological progress than would have obtained under a system that did try to reward independent inventors.

In response, however, it should be noted that the extent to which the patent system benefits society is not merely a function of how much technological innovation there is, but at what social cost. If research efforts are not skillfully conducted, or if they are skillfully conducted but wastefully duplicative of one another, then the net benefit to society is diluted. In this connection, then, the policy of not considering claims to independent-but-subsequent invention might be defended on the ground that it is appropriately cost reducing. Those who are relatively less well qualified to undertake research and development are discouraged from entering the race in the first place, and those who are well qualified and well equipped are encouraged to proceed with deliberate speed. Moreover, although duplication of research effort cannot be altogether eliminated without sacrificing the stimulating effects of free and open competition, it is plausible to suppose that the more expeditiously a technological advance is brought forth and placed on the public record, the less time others will be wasting independently trying to invent it.

The Contrast between Copyright and Patent

Before concluding this survey of possible justifications for the system of intellectual property rights, it would be worthwhile to consider a striking contrast between copyright and patent law. An invention is not patentable unless it is nonobviously new; the fact that someone else subsequently arrives at it on his own, without unfairly free riding upon the efforts of its first discoverer is, as I have shown, given no weight. To obtain copyright protection, however, a putative author need not demonstrate that the work in question is unobviously new and different. Instead of "novelty" and "nonobviousness," all that matters is "originality"—where this means, quite narrowly, that the work in question independently originated with the copyright claimant.

Thus, someone who through his own independent effort arrives at a work virtually identical to an already copyrighted work does not infringe upon the rights of the first copyright holder. Indeed, that person is entitled to copyright protection of his own as well. If, in contrast with patent holders, copyright holders do not have the right to exclude a strikingly similar result of independent origin, then what protection do they have? Quite simply, they have the right not to have their work copied, reproduced, or adapted. (Depending on the nature of the work or the medium in which it is fixed, they also have exclusive right to its distribution, display, or performance.)

For example, if a work of factual information—such as a map, a directory, or a statistical chart—has been independently created, then despite the fact that it is virtually identical to an already copyrighted work, it does not count as a case of copyright infringement. The putative author need only show that he went to the trouble of gathering and recording the information on his own. A further illustration of this point is provided by the case of someone who reproduces a work of art that is no longer under copyright protection. Accurately reproducing a work of art involves effort and expense. The copyright system allows someone who has taken the trouble of reproducing a work of art to obtain copyright protection (see Alfred Bell & Co. v. Catalda Fine Arts, Inc., 191 F.2d 99[2d Cir. 1951]).

What this means, however, is that without the copyright holder's permission, nobody has the right to copy or reproduce his reproduction. Anybody else who wishes to market reproductions of the work must go through the process of independently reproducing the original work of art. If the finished product turns out to be virtually identical to the copyrighted reproduction, the original copyright is not thereby infringed.

Can this apparent disparity between the way in which the patent system treats independent but obviously similar inventions and the way in which the copyright system treats original though not necessarily novel works of authorship be explained? More important perhaps, can it be justified? Of course, these two bodies of intellectual property law have different domains. Patent law deals with utilitarian works falling into several well-defined categories (machines, compositions of matter, and so on), whereas copyright law deals with the nonutilitarian aspect of virtually any work of original human authorship, provided that it can be fixed in a tangible medium of expression.

Perhaps the difference in the way the two systems handle similar but independently arrived at results can be explained by reference to the difference in their respective subject matters—that is, by the fact that in the one domain the result is an invention and in the other a work of authorship. The examples cited above suggest that in contrast with the patent system, the copyright system is a bit less concerned to provide the maximum incentive to the creation of that which is new and nonobviously different and somewhat more concerned to prevent unfair free riding. If this were really true—if these two systems of intellectual property law did assign different weights to these two concerns—would the difference between authorship and invention shed any light on why they do?

It is possible, of course, that this particular difference between these two spheres of intellectual property law is nothing more than a historical accident, something that can be explained perhaps but not reasonably justified. On this view, a more rational intellectual property system would have eliminated the disparity, either by making independently arrived at inventions patentable or, more probably, by denying copyright protection to independently authored works that are strikingly similar to works already copyrighted.

Although a full assessment of this issue would be beyond the scope of the present discussion, I believe that in this particular respect, the present system is not unreasonable. To see how it might be justified, I focus attention on the three concerns whose special relevance has been established by much of what I have considered up to this point: the protection of basic freedom of thought and expression; the encouragement of authorship and invention; and the prevention of unfair free riding.

Recall that two different lines of argument seemed to support the conclusion that the patent system should not attempt to honor claims to independent invention. One, an effort to make the system fairer in this respect is likely to generate more, rather than less, unfairness. And two, in a system that attempts to verify and honor such claims, there is likely to be more time-consuming and costly litigation, less expeditious research effort, and more wasteful duplication. But do the same arguments apply, and with the same force, to the copyright question? I suggest that they do not. What follows is merely a sketch of how one might prove the point.

Many copyrightable works are primarily products of the creative imagination (novels, plays, poems, musical compositions). With works of this kind, strikingly similar surface expression is strong evidence that the work that comes later is not of genuinely independent origin. Thus, a copyright system that seeks to reward genuinely independent authorship in this category of works is not so liable to be subverted by fraud and deception as a patent system that seeks to reward independent-but-subsequent invention. The attempt to make the system fairer in this respect does not promise to generate more unfairness than it alleviates. Nor does the social cost in effectively litigating the cases that might actually arise threaten to be excessive.

The fact remains, however, that there are many copyrightable works whose primary purpose is to convey information. They are formed by an attempt to gather and record what is already out there, as well as by certain conventions of information presentation (for example, a directory will generally place its listings in alphabetical order). It is therefore not entirely improbable that these works (or significant portions thereof) will be identical to one another, not only in their informational content, but even in the particular expression that they give to that content. With such works there does seem to be more opportunity for successful fraud and deception. Thus, for example, it is relatively easy to copy someone else's map or statistical chart and then, making a few cosmetic changes, pass it off as the product of one's own research.

Even in this general area of copyrightable works, those who wish to maintain their copyright protection have certain strategies at their disposal. The authors of a question-and-answer game such as Trivial Pursuit have the option of deliberately planting a few incorrect answers to help establish whether other game vendors are simply stealing their materials. In the same spirit, cartographers can place a few relatively insignificant but inaccurate details in the maps they sell in order to help them detect free-riding competitors. Even so, such devices can be used only sparingly or else the informational value of the work will be compromised. And in any event, it is plausible to believe that under a policy that entertains the defense of independent origin, there will inevitably be at least some undetected fraud and deception that would not have occurred otherwise. Why then does the copyright system not operate, at least in respect to works that are primarily information disseminating, in a manner more parallel to the patent system?

Here is a point at which one might do well to recall the third pillar of the intellectual property system: the concern to protect basic freedom of thought and expression. One of the virtues of a social system in which there is freedom of thought and expression is that published factual claims can be subjected to independent scrutiny and testing, and then—depending on the results—be corrected or corroborated. If the original claims were factually accurate, those who undergo the rigors of independently researching the same matter and disseminating their findings may well discover that their efforts have culminated in a finished product virtually identical in some respects to portions of an already copyrighted work. If, however, the rules of the system were altered so as to classify such people as copyright infringers, there would be a strong disincentive to independent research. This, in turn, would be to overlook an important aspect of the free-speech system: society's interest in making progress toward greater accuracy of publicly available information.

Concluding Remarks

I began by asking whether the intellectual property system as we know it confers ownership of ideas. In arriving at a negative answer, I also came to the realization that an important constraint operating upon the design of the intellectual property system is the concern to preserve basic freedom of thought and expression. Patents and copyrights give people special rights, not to ideas as such, but to their practical application and to their particular expression. I then investigated possible justifications for instituting the rules of the patent system. Some arguments (from personal liberty, from the right to privacy together with the right to make contracts) proved too little. Other arguments (from an alleged right to the products of one's own mental activity, or from just desert) would, if they were to work, prove far too much. They also, as it happened, proved to be inherently confused and implausible. This left two reasonably plausible and relevant concerns: to promote technological progress and to prevent unfair free riding.

After exploring a number of complications that are likely to attend the application of these considerations to the design or reform of the intellectual property system, I paid special attention to one particular feature of the patent system: the fact that those who independently but subsequently arrive at a patentable invention do not share, with its first discoverer, the privileges and prerogatives of the system. At first, it seemed as though this feature might warrant the view that prevention of unfair free riding is not a genuinely independent concern of the patent system or that the system is unfair. I then showed how two arguments—one from fairness itself, and the other from the need to provide an efficient incentive to technological progress—combine to provide reasonably strong support for what might otherwise seem to be a problematical feature of the system.

To conclude my discussion, I considered an important contrast between copyright and patent law. Whereas patent holders can exclude even those who have independently arrived at what is virtually the same invention, copyright holders do not have the right to exclude a virtually identical work of genuinely independent origin. This contrast might lead one to suppose that in the realm of copyright, prevention of unfair free riding is (for some unexplained reason) a much weightier consideration than the encouragement of nonobviously different works. I then explained how this appearance might be misleading: first, because the arguments (from fairness and from incentive to progress), which justified the policy of not permitting independently arrived at inventions to be patented, do not lead to the parallel conclusion with respect to a significant range of copyrightable works; and second, because an additional consideration—the need to protect the system of free thought and expression and some of the values embodied in that system—also has a special role to play in shaping this particular feature of copyright law.

Indeed, if one very general conclusion has emerged from all the many more specific issues I have discussed it is just this: various features of the intellectual property system will be more or less justifiable to the extent that they reflect a satisfactory balancing of three major concerns: freedom of thought and expression, incentive to authorship and invention, and fairness.

Acknowledgment

I received valuable criticism on earlier versions of this essay from the Cal Tech Philosophy Discussion Groups, the Society for Philosophy and Public Policy in New York, and in particular, Stefan Mengelberg, a member of the latter group.

References

Merton, R. K. 1967. *On the Shoulders of Giants: A Shandean Postscript*. New York: Harcourt Brace Jovanovich.

Nozick, R. 1974. *Anarchy, State, and Utopia*. New York: Basic Books.

Rand, A. 1967. *Patents and Copyrights in Capitalism: The Unknown Ideal*. New York: New American Library, p. 133.

Rothbard, M. N. 1977. *Power and Market: Government and the Economy*. 2d ed. New York: New York University Press, p. 71.

12. Patents, Natural Rights, and Natural Property

Michael Davis

One thesis of Arthur Kuflik's "Moral Foundations of Intellectual Property Rights" is that patent law as we know it is inconsistent with a natural right to property. For example, he points out that patent law does not treat as equals all those who independently hit upon the same invention (Kuflik this volume). While accepting most of what Kuflik has to say, I shall argue that, on the contrary, patent law is consistent with a natural right to property.

The theory on which my argument depends, though unusual and rather weak, is more plausible than those to which theorists commonly give pride of place. The resulting right does not, however, undermine Kuflik's ultimate thesis, that there is no natural property in one's discoveries or inventions. Like Kuflik, I shall begin with some puzzles.

First Puzzle: The Constitution and the Fifth Amendment

Our thinking about a natural right to property seems to assume a connection between that right and the state of nature. To say someone has a natural right to property is (it seems) to say, in part at least, that such a right is more or less independent of law, that it is a moral right that could exist in nature as well as in civil society, that there is *natural property*. Such property is a claim on government (once government comes into being) just as my right to life or liberty is a claim that government needs very good reason to ignore. The right need not be absolute. Government may be justified in seizing my property to pay a just debt I owe, to compensate another for the loss I wrongly caused him, to exact just punishment, or otherwise to correct injustice. But any other taking of my property, however justified, must leave behind a debt. If necessity requires taking without compensation, the taking may be excused but it cannot be justified. Government cannot rightfully take my property for any use, however justified, without compensating me (or getting my personal consent).

This, roughly, is the natural right to property that Kuflik assumes. He certainly has reason for that assumption. Robert Nozick's *Anarchy, State, and Utopia* (1974) has given it wide currency. One can also read it into documents written by those commonly supposed to believe in a natural right to property. The best example for my purpose is the U.S. Constitution (with its original ten amendments). While the Constitution itself says nothing about a natural right to property, the Fifth Amendment seems to protect property in much the way the theory of natural property would lead one to expect. Congress is forbidden to take property "in a criminal case . . . without due process of law" or "for public use without just compensation." The silence of the Constitution (and later case law) seems to imply that Congress has no power at all to take private property merely for private use.[1] The Constitution does, of course, allow Congress to tax, and for many libertarians that seems to be a taking for public (or private) use without just compensation. But, one can, I think, see how the services provided with the money raised by taxation might be construed as adequate compensation—provided, at least, the purpose is not primarily to redistribute wealth but to provide for the common defense, promote the general welfare, or the like.

My first puzzle concerns what happens when this theory of natural property is set beside the only provision of the Constitution relevant to patents. Clause 8 of article 1, section 8 sounds like it was written by someone who did not accept the theory. The Constitution simply grants Congress the *power* to "promote the progress of science and useful arts by securing for limited times to authors and inventors the exclusive right to their respective writings and discoveries." The emphasis is on promoting progress in science and useful arts, not on securing fundamental rights. The amount of protection is left open (except that it must be for a "limited" time). Indeed, the Constitution does not expressly require a patenting system. Congress is granted a power—that is all. Congress can either exercise that power (permitting the granting of exclusive rights for a limited time) or not exercise it (leaving inventors to use secrecy or whatever other methods they legally can to control their discoveries). Presumably, Congress even has the power to decide tomorrow to abolish the patent system. If, on this reading, there is any legal limit to what Congress can do with patents, it is that Congress could not terminate existing patents early without just compensation. Such a termination, even if justified by some public purpose, might, it seems, still be a taking for which the Fifth Amendment could require compensation.

But, if the authors of the Constitution believed in natural property, should they not have required Congress to protect all property, even intellectual property, by some reasonable means, instead of leaving such protection to the discretion of Congress? Well, perhaps not. As Kuflik makes wonderfully clear, patents (as we know them) seem inconsistent with a natural right to property as it is understood, and with other natural rights, too. A patent not only guarantees an inventor control over his own discoveries—the products of his own labor—it also bars others from benefiting from their wholly independent discoveries—the products of their own labor.[2] And it does that without compensating them for their labor or lost opportunity. Independent inventors are in that respect worse off with a patent system than without it. If natural law gives one a right to the products of one's own labor, then granting a patent must sometimes take another's property without just compensation—that is, take an independent inventor's right to benefit from the otherwise salable product of her own labor.

An ingenuous natural-rights theorist might try to get around this embarrassment by claiming that each such inventor is compensated by the system as a whole. The same system that took the right to her invention because she was not the first to hit upon it gave her in return the right to exclusive control had she been first. If these rights are of relatively equal value, the system has compensated her already. The argument may be strengthened by pointing to the difficulties of verification of independent discovery, the general benefits derived from such a spur to break new ground, and so on.

But such ingenuity will not solve the puzzle that concerns me. What puzzles me is not the particular difficulties of making the natural-rights argument work but the obvious incongruence between the way of thinking such a theory seems to require and the way those supposedly holding the theory actually think about patents. Consider, for example, comments that Thomas Jefferson made in 1813 on the thesis that "inventors have a natural and exclusive right to their inventions":

> While it is a moot question whether the origin of any kind of property is derived from nature at all, it would be singular to admit a natural and even a hereditary right to inventors . . . Stable ownership is the gift of social law and is given late in the progress of society. It would be curious then if an idea, the fugitive fermentation of an individual brain, could, of natural right, be claimed in exclusive and stable property. If nature has made any one thing less susceptible than all others to exclusive property, it is the action of the thinking power called an idea. (Dumbauld 1955)

These comments were made by the same Jefferson who only a paragraph earlier had declared that laws "abridging the natural rights of the citizen should be restrained by rigorous constructions within their narrowest limits." Jefferson is not in doubt about natural rights as such. His doubts concern "the origin of . . . property . . . from nature," especially the very possibility of natural property in inventions. He expressly declares property "the gift of social law." His comments appear to fit the constitutional provision for patents (though he opposed that provision) and the Fifth Amendment (which he supported) better than the theory of natural property Kuflik criticizes.[3]

What is one to make of this? One possibility is that Jefferson is an anomaly. I tentatively dismiss that possibility for lack of evidence. My reading of other Founding Fathers leaves me with the distinct impression that their views on natural property were much like his.[4] Another possibility is that the framers of the Constitution were generally confused on this subject. I think that that possibility should be dismissed, too. Given the fit between what they said and what they did, it seems extravagant to claim that what they really had in mind was something else. So, if the Fifth Amendment is to be understood as protecting a natural right to property, it seems that it must be a natural right that does not require natural property. What kind of natural right is that? I shall try to answer that question, but first another puzzle.

Patent Law

Looking at the patent law itself, one finds distinctions having a consequentialist justification not likely to override a claim of natural right. I shall focus on some distinctions Kuflik also discusses. One may not patent laws of nature, mathematical formulas, or methods of doing business, even if the discovery is of something nonobvious and useful. But one may patent compositions of matter, machines, articles of manufacture, methods of making something, and methods of using something, provided they are useful, nonobvious, and not already patented. One can, I think, find a nonconsequentialist principle for not allowing patents on laws of nature or mathematical formulas. These are learned, one might say, not invented. They are not the work of their discoverers—though, of course, their discovery is. One might also find a nonconsequentialist principle behind allowing the patenting of machines, articles of commerce, and various methods of making something. These are invented by someone, not just learned. They are the product of his labor, hence his. But, by this reasoning, compositions of matter should not be patentable (at least when they can, like most drugs, be found in nature) while methods of doing business should be patentable.[5]

One response to this nit-picking is to agree. It might be said that methods of doing business, if useful and nonobvious, should be patentable and that a mere "composition of matter" (for example, a naturally occurring drug like penicillin) should not be patentable, whereas the method of producing it should be. This response would certainly make it easier to find a justification for patent law based on natural rights (though many of the other problems Kuflik points to would remain). I am nevertheless unhappy with it. Let me explain.

Some work is such that, if not regulated, the balance between initial investment and probable return will be so unfavorable that no one could rationally undertake the work from ordinary business motives. If everyone would benefit from the work being done, no matter who did it, and if the work would not be done unless done from ordinary business motives, the existence of such work poses a problem of social coordination to which governmental regulation is a solution. Regulation can change the probable return so that the work can be rationally undertaken from ordinary business motives.

Patents provide such regulation. For example, the promise of a patent can make financially attractive an expensive program of research and development even though the outcome will be a product that others could easily and profitably copy. Patents can do that by assuring the would-be inventor that he need not fear such copying. Those taking advantage of his efforts will not get a free ride. They will have to pay him substantially more than that least-profitable of compliments, imitation.

This description of the patent system may suggest that patenting can be justified by the principle of fairness. Those who voluntarily benefit from a just practice (by copying the inventions or discoveries that would not have been made but for the patent system) should (it might be said) take on a fair share of its burdens (the restraints and payments the system requires). But the appeal to fairness itself presupposes that the patent system is relatively just. And that is exactly what Kuflik's discussion of natural rights makes doubtful. If patent law really does involve taking the natural property (or liberty) of others (that is, of independent inventors), then patenting is at least prima facie morally wrong. Since mere convenience should not trump natural rights, it is hard to see how the system could be justified.

After all, though patents provide a solution to the problem of social coordination that I am considering, they are not the only possible solution. Government can use its power of taxation to support research by direct grants to researchers, by establishing research institutes of the appropriate sort, or by making research the business of this or that government department (as it has done for agricultural research). Government can also use its taxing power less directly—for example, by exempting from taxation certain kinds of research or gifts in support of such research (as it does, for example, by exempting universities from taxation). Where government has chosen to encourage research by allowing patents, it has done so because that seemed the best way to encourage research, not because it seemed the only way. Where then does natural property fit into the justification of patent law?

Here the modern problem of *orphan drugs* is suggestive. These are drugs that have been known for a long time (and so, have no research costs). They are orphans because, without an exclusive right to make them, no one will make them. There is a problem of social coordination. For example, manufacturers may believe that, given the substantial investment necessary to produce a certain drug, the market for it is too small to support more than one manufacturer and that, given the relative secrecy in which manufacturers sometimes prepare drugs for production (and the long lead time), there is no way to know that another manufacturer is not already preparing to produce the drug. Absent the grant to some manufacturer of an exclusive right to make the drug, government will itself have to make it, subsidize its private manufacture, or let those who need it do without.

The justification of granting patentlike protection to such drugs seems to be much the same as granting patent protections to newly discovered compositions of matter (rather than merely to some method of manufacturing them). Society will benefit from the drug's manufacture. That such a drug was discovered long ago (and *not* by the applicant for patentlike protection), seems not to weaken (or strengthen) the argument significantly. One can, however, see more clearly than for patents why regulation should exclude all others from the inventor's work. That exclusivity is a practical solution to a practical problem.[6]

So here is my second puzzle—a system of patent law that seems easier to justify without a theory of natural property than with it.

Solving the Puzzles

What these two puzzles suggest to me is that people may have misunderstood what is meant by a natural right to property. I began by supposing that such a natural right required *natural property*, a moral right over something that might exist even if there never had been government. But I have found the Constitution and patent law seeming to take for granted that civil society can create property—that (as Jefferson put it) an exclusive right over an idea is "the gift of social law." While that finding is (as Kuflik claims) inconsistent with natural property, it is nevertheless consistent with a more limited but still natural right to property. Taking the Fifth Amendment as the text, one might claim that a natural right to property means that, once someone owns something, she should not be deprived of it except with her consent, to correct an injustice she did, or for a public purpose when, taking into account all factors including payments to her, her condition is not made worse by the taking. One has a natural right to what has been legally granted to her, not a natural right to such a grant.

This is not, I think, among the classic theories of property. I cannot think of any political theorist who has embraced anything much like it (though it does seem like the sort of theory that Lon Fuller [1969] should have embraced).[7] Those theorists who (like Hobbes or Rousseau) think of property rights as conventions think that the "sovereign" can justly abolish property at will without compensation, while those who (like Nozick) think that the "sovereign" cannot justly take property without paying compensation (and maybe not even then) also seem to think of property as existing more or less independent of actual law. Locke, though holding a theory of natural property, seems to allow the majority to take property without compensation.[8]

Yet, the theory suggested by this consideration of patent law seems more attractive than any of these. On the one hand, it permits us to give full weight to consequences, desert, and the like when considering whether to have a certain sort of property. We are not subject to any tyranny of natural rights. On the other hand, once we have a certain sort of property, the theory forces us to recognition of claims of actual proprietors to what they have. The rights they have are now claims on government independent of the future of the system that granted them. The rights are natural in the sense that they constitute claims not extinguishable by law, even though the law created them.

Conclusions

Though I have been speaking till now of patent law, I must admit that other forms of property seem to me to differ from patents in no respect relevant here. All can, I think, be understood as the solution of this or that problem of social coordination. Indeed, that is exactly what Hobbes did for "meun and tuum" generally and Locke for private ownership of land and goods in the state of nature. The Founding Fathers also do not seem averse to generalizing the analysis in this way. The Fifth Amendment is concerned with all property, not just with intellectual property. The Constitution nowhere expressly gives greater protection to rights in land, goods, or the like. Indeed, there is even good reason that the Founding Fathers should not be averse. The limited natural right to property I have hit upon seems capable of protecting everything usually in danger when people set their natural right to property against what some government is proposing. That is, the right to property is not typically invoked to demand legal recognition of a new form of property or to prevent abolition of an old one, but to prevent *confiscation* of existing property. The limited natural right to property also seems easier to defend than any right presupposing natural property. No theory of natural property is needed. The limited right not to be deprived of what the law has given seems to rest upon nothing more controversial than a moral rule like, Do not harm others. Taking without just compensation something a person already has seems to be prima facie morally wrong in a way that failing to make it legally possible for him to have it in the first place does not.

So, my peregrinations in a distant province of the law have brought me to an object of more than pedestrian interest: a theory of natural right to property both more defensible than the received theory and more consistent with patent law as we know it. But, I have found that theory at some cost. This best theory provides absolutely no support for the claim that inventors have a natural property in their inventions. I have ended up within hailing distance of Kuflik.

Notes

1. See, for example, Pennsylvania Coal Co. v. Mahon, 260 U.S. 393, 415 (1922):

> The protection of private property in the Fifth Amendment presupposes that it is wanted for public use, but provides that it shall not be taken for such use

> without compensation. . . . When this seemingly absolute protection is found to be qualified by the police power, the natural tendency of human nature is to extend the qualifications more and more until at last private property disappears. But that cannot be accomplished in this way under the Constitution.

Cf. Justice Rehnquist's dissent in Penn Central Transportation Co. v. New York City, 438 U.S. 104, 138 (1978).

2. I have chosen as exemplar "the labor theory of property" rather than some other theory of original acquisition primarily because the labor theory seems to fit patent law better than any alternative. The two most promising alternatives, "first occupation" and "desert," would make my puzzle even harder to solve. Both seem to entail, as the labor theory does not, that the discoverer of a law of nature or mathematical formula should have the same right to patent as the discoverer of a composition of matter or inventor of an article of manufacture.

3. The following appears in a letter from Thomas Jefferson to James Madison dated 31 July 1788:

> The saying there shall be no monopolies lessens the incitements to ingenuity, which is spurred on by the hope of a monopoly for a limited time, as of fourteen years, but the benefit even of limited monopolies is too doubtful to be opposed to that of general suppression [of patenting]. (Dumbauld 1955, pp. 141– 142)

4. See, for example, John Adams's Massachusetts Constitution of 1779, which reads:

> Each individual of the society has a right to be protected by it in the enjoyment of his life, liberty, and property, according to standing laws. He is obliged, consequently, to contribute his share to the expense of this protection. . . . But no part of the property of any individual can, with justice, be taken from him or applied to public use without his own consent or that of the representative body of the people. (Peek 1954)

Note the absence of any provision for compensation.

5. I have avoided Locke's metaphor of "mixing one's labor" to describe acquisition because, whatever its merits for thinking about gaining a property in particular objects (for example, an apple or piece of ground), it seems hopelessly strained when the concern is a type of thing (for example, a composition of matter or method of making something). For an axiomatization of Locke's labor theory that does without the mixing metaphor, see my article in *Ethics* (1987).

6. A similar defense was made of early copyrighting. For example, a limited monopoly over the publication of the Bible would be defended as necessary to prevent "disorder" in the market, which might lead to no one's printing it even though it lay in the public domain. See, for example, Patterson (1968).

7. While I have found no theorist who offers such a theory, I have found one nontheorist who seems to have come pretty close to an explicit expression of it. During the debate between Cromwell's officers and the Levellers in 1647, General Ireton made a number of comments like the following:

> Divine law extends not to particular things . . . [If] a man were to demonstrate his right to property by divine law [that is, under the commandment, "Thou shalt not steal"], it would be very remote. Our right to property descends from other things [that is, particular positive laws], as well as our right of sending burgesses. (Woodhouse 1951)

Ireton was, of course, responsing to the Levellers who held a theory of natural property (as well as of a natural right to vote). For the most relevant discussion, see Woodhouse (1951, pp. 53–63).

8. In his *Second Treatise of Civil Government*, Locke observes the following:

> Men . . . have such right to the goods which by the law of the community are theirs, that nobody hath a right to take their substance or any part of it from them without their consent. . . . Hence it is a mistake to think that the supreme or legislative power of any commonwealth can do what it will, and dispose of the estates of the subject arbitrarily, or take any part of them at pleasure. . . . But it is true, government cannot be supported without great charge, and it is fit every one who enjoys his share of the protection should pay out of his estate his proportion for the maintenace of it. But still it must be with his own consent—i.e., the consent of the majority, giving it either by themselves or their representatives chosen by them. (Cook 1947)

Although these passages can be read as implying the theory of natural right to property that I have been suggesting, the more common interpretation is contrary. For Locke, the opposite of "arbitrarily" is "according to standing laws"; hence, so long as the legislature took by standing law, not individual decree, it could, it seems, take without compensating an owner. Cf. the language of the Massachusetts Constitution in note 4, above.

References

Cook, T. I., ed. 1947. *Two Treatises of Government*. New York: Hafner Publishing, secs. 138, 140.

Davis, M. 1987. "Nozick's Argument for the Legitimacy of the Welfare State." *Ethics* 97: 576–594.

Dumbauld, E., ed. 1955. *The Political Writings of Thomas Jefferson*. New York: Liberal Arts Press, p. 56.

Fuller, L. L. 1969. *The Morality of Law*. New Haven: Yale University Press.

Nozick, R. 1974. *Anarchy, State, and Utopia*. New York: Basic Books.

Patterson, L. R. 1968. *Copyright in Historical Perspective*. Nashville: Vanderbilt University Press.

Peek, G. A., ed. 1954. *The Political Writings of John Adams*. Indianapolis: Bobbs-Merrill, pp. 97–98.

Woodhouse, A.S.P., ed. 1951. *Puritanism and Liberty*. Chicago: University of Chicago Press, p. 60.

13. Commentary: Legal and Ethical Issues

Gerald Dworkin

In this commentary, I will focus on some of the areas of agreement and disagreement among participants in a conference on intellectual property held in late 1985 (National Science Foundation 1985). I will pursue some of the implications of this agreement and disagreement.

Consensus

There has been general agreement that the specific forms of protection for intellectual property (that is, copyright and patents) are to be understood and valued as *means* rather than *ends*. Such agreement limits the possible candidates for justifying or criticizing these institutions. Thus, for example, it is unlikely that notions of desert or entitlement to the fruits of one's intellectual labors would be an important justificatory base. This is because, for all but the crudest forms of utilitarianism, the value of rewarding desert or satisfying entitlements is not purely instrumental but is desirable for its own sake. Indeed, if one thinks of, say, reward for effort, then patent law (which rewards not merely effort but effort plus being first to file) seems particularly inappropriate. One must consider those industrious souls who labored intensively but unfortunately did not make the discovery or invention. Surely, if effort is important, then they deserve as much as the successful innovator.

There are both general and specific difficulties with an alternative justification for intellectual property based on an inventor's claim to a natural right over the fruits of his or her labor. A convincing argument has never been made for assuming that when, as Locke would put it, we "mix" our labor with some natural resource, we gain the right to exclusive possession of the product (Cook 1947). Why do we not, as Robert Nozick has slyly asked, instead lose our right to the labor we have mixed into the product (Nozick 1974)? For intellectual property, there is no tangible resource—such as a piece of wood or plot of ground—about whose possession we can raise questions of ownership. If I have a new idea about how to manufacture widgets, then no one can take my idea from me—although they can use it. The notion that I can prevent others from using my ideas must be weighed against the possibility that others will independently arrive at my idea after I did and will be prevented from using it because I came up with it first.

If we concede that patents are means, then the justification will be consequentialist in nature—that is, judged by success in promoting certain ends. Of course, there may be evaluative considerations relating to which means are proper to attain the ends, and thus there is still room for considerations of entitlements, desert, and so forth.

Dissensus

There is apparently great disagreement with respect to two central issues in intellectual property policy: the relation of the means to certain ends, and the nature of the ends themselves.

I am particularly struck by the absence of evidence for the proposition that patent or copyright effectively promoted (or promoted more effectively than alternative schemes of protection) even the least controversial end—the promotion and protection of technological innovation. Some have actually used the word *faith* in the context of assessing the view that patents indeed worked. One piece of concrete evidence—a survey by economist Sidney Winter—indicates that those individuals who have the greatest stake in the patent system seemed to think that it does not play an important role relative to other considerations (Winter this volume).

It is certainly difficult to test the proposition that we would be better or worse off in terms of the amount and nature of technological innovation if we abandoned or altered the existing system of protection for intellectual property. But the difficulty is no warrant for making unsupported claims about the efficacy of patents or copyrights. I am always skeptical of a priori psychologizing about what motivates people to produce or create. It is popularly asserted that inventors, artists, and writers would not engage in their creative activities unless they could exclude others from the enjoyment of their products without payment; this seems to me a wholly unsupported generalization. It may be true that those who seek oil would not do so absent some restriction on the use of what they find, but I would guess that neither the Wright brothers, nor Beethoven, nor Kafka would have behaved any differently had there been no protection on the use of their works by others. We need to see more works and less faith.

Another area of great disagreement over intellectual property policy is the definition of the ends to be promoted by the system of intellectual property rights. It is obvious that technological innovation is only one possible end in which people are interested. The following ends (among others) were mentioned at least once at the aforementioned conference:

1. More money for universities
2. Improvements in public health
3. Securing openness of communication within the scientific community
4. Preserving the values of the academy

5. Maximizing wealth
6. Keeping options open
7. Strengthening the public sector
8. Promoting a sense of social responsibility among scientists for the consequences of their research
9. Ensuring quality control for new products, particularly drugs
10. Supporting the development of an information society

This multiplicity of ends complicates the discussion in at least two ways. First, the empirical problems become very complex. We must do counterfactual history about the causal relationships of intellectual property to a system of multiple ends. Second, we have to think about trade-offs among the ends, ways of assigning weights to the value of these ends. Because it is unlikely that we shall reach any consensus about such a system of trade-offs, we must also make decisions about the appropriate institutional forums for reaching authoritative decisions about which ends to pursue, and must do so in light of the best evidence we have about what means can most efficiently and justly promote those ends.

As the essays in this volume show, the problems are more likely to be political and legal than distinctively moral in character. Buttel and Belsky are interested in how power is distributed between corporations and the government. University faculty are worried about how patenting will affect the governance of universities. Lawyers like Samuelson are worried about how the legal system should best adapt to changing circumstances that could not have been envisaged when the original constitutional provisions for patenting were established. In denying that the problems are particularly moral in character, I am not denying that they are important, that they are serious, or that they have dimensions larger than the purely technical. Rather, I am suggesting that the interpersonal dimension is not likely to be significant and that, therefore, most traditional moral theories (with the exception of utilitarianism—which is, I believe, a political rather than a moral theory) are unlikely to give much guidance in these matters.

References

Cook, T. I., ed. 1947. *Two Treatises of Government*. New York: Hafner Publishing.

National Science Foundation. 1985. Conference on Ethical Implications of Trade Secrecy, Patents, and Related Property Controls for Science and Technology, Illinois Institute of Technology, Chicago. EVIST Grant No. RII-830-9873.

Nozick, R. 1974. *Anarchy, State, and Utopia*. New York: Basic Books.

14. The University, Scientific Research, and the Ownership of Knowledge

Leonard G. Boonin

This chapter is divided into three sections. The first contains a general analysis of the concept of ownership of knowledge and the philosophical basis for recognizing property rights in knowledge. The second examines the legal basis of such rights in the United States. The final section explores, in the light of that background, the conflict that has arisen within universities between the values underlying scientific research and the legal norms governing intellectual property.

Philosophical Basis

The notion of owning knowledge is somewhat ethereal and until recently had little relevance to academic institutions. Outside the formal context of the law of copyright, patents, and trade secrets, we are not likely to speak of ourselves as owning or having a proprietary interest in knowledge. It is much more natural for us to say that we possess certain knowledge than that we own it. We may manifest such possession in a variety of ways, such as by communicating it to others or employing it in what we do and make. We may also decide not to divulge or make use of what we know. I shall later show the significance of a right of nondisclosure in understanding the concept of ownership of knowledge.

While possession and ownership are related concepts, they are clearly different. *Possession* refers to having control, while *ownership* refers to having rights to control. We understand the difference between a thief who has physical possession of a car and the rightful owner who may never see it again. Ownership is a normative concept that implies a complex set of rights. What is included in that set may vary depending on the nature of the object involved. It may include the right to use, sell, rent, give away, abandon, consume, or even destroy. Roughly speaking, those rights reduce to two different kinds: positive rights of access and beneficial use and negative rights to exclude others from its use without permission.

Sometimes that bundle of rights is divided in ways that make it difficult to say precisely who the owner is. You may, for example, receive under a will the right exclusively to possess and to use an object for the length of your life with provision made as to how it is to be disposed upon your death. No one else alive may have any rights in relation to that object, yet you do not fully own it.

Ownership may roughly be characterized as possessing the full set of rights a particular object lends itself to or, if that bundle is divided, having the residual rights remaining after all other proprietary interests have expired. Thus a landlord is the owner of a building even though the tenant has an exclusive right of possession for a specific length of time. In some cases, such as with ninety-nine-year leases, the possession of a legally limited proprietary interest may be economically more valuable than the reversionary rights held by the owner. The distinction between owning property and having some more limited property interest will come into play when I examine how the legal system approaches intellectual property.

In principle, any object that is capable of being controlled is capable of being owned. We own discrete physical objects—for example, land and personal property—to which access is easily controlled. Some highly mobile physical substances—such as air and stream water—are almost impossible to individuate or control in this manner, and the concept of ownership becomes problematical. A farmer may be entitled to a certain quantum of water from a navigable stream adjacent to his or her property but not literally be the owner of some determinate body of water.

The most basic control we have over our knowledge concerns decisions about whether to communicate it. If we do communicate what we know, the subsequent movement of that knowledge is determined by the behavior of those to whom we make our communication. We may seek to limit its access by imposing conditions of confidentiality, but there are clearly risks that it will be further dispersed in ways not authorized or contemplated. In addition to directly communicating what we know, we make use of our knowledge in what we do and make. In some cases others are able to decipher the knowledge that provided the basis of our activities. Of course, we have gained most of our knowledge from others and hence we have little control over its dissemination or the use made of it. Most knowledge is in this sense in the public domain. Still there can be important kinds of information or knowledge that only we possess.

One of the peculiarities of knowledge as property is precisely that it is capable of being fully possessed at the same time by an indefinitely large number of persons. Knowledge also differs from other objects in not being capable, in any ordinary sense, of abandonment or destruction. Nor can exclusive possession be easily regained once it has been communicated. Even if a person never discloses to others certain information, he or she may not be completely assured of having exclusive possession if there are other ways of gaining access to it.

Sometimes the value of knowledge is dependent on its being widely shared. This is especially so in cases requiring cooperation and coordination among persons—for example, the importance of everyone's having innoculations, or following the rules of the road in driving. In competitive situations, on the other hand, it may be in our interest to exclusively possess knowledge—for instance, that a painting being sold at auction is an authentic Rembrandt, or the formula for making a popular soft drink. In still other situations, knowledge may become a valuable commodity for which others are willing to pay—such as a physician's knowledge of how to treat an illness.

The decision to recognize certain things as property is a function not only of the degree of difficulty involved in controlling them, but of the degree of value and scarcity they possess. The easier it is to exercise control, and the greater its perceived value and scarcity, the more likely it will be recognized by society as constituting property and as being protected by social and legal norms. In deciding to provide legal protection, a society has also to take into account the administrative costs of establishing reliable legal standards and implementing them.

That society judges something to be worthy of being socially or legally respected as property does not answer the question of who should have what specific rights in relation to the object in question. Things can be owned collectively by society as a whole, by particular groups or entities within society, or by individuals. In a society with collective ownership, the rights of individuals and groups will mainly be rights to access and use. If a community owns something collectively, it claims the right to exclude nonmembers from access, as well as the authority to regulate the access by members. Under a system of private ownership, it is the individuals or groups within the community that are recognized as having the authority to determine who has access and is entitled to beneficial use.

What forms of knowledge ought to be recognized as constituting property, and who should have proprietary rights in it? Among the factors relevant in establishing social policy are how difficult and costly it was to acquire the knowledge and how important it is that the knowledge be shared. The more difficult and expensive it is to acquire certain kinds of knowledge, the more it can be argued that the possessor of such knowledge is entitled to receive some reward or compensation for disclosing it. The more important it is for others to obtain that knowledge, the more grounds there are to impose a duty on those who posses it to share freely.

There are some forms of important information to which people generally have easy access, and which are shared freely. We learn a lot of detailed information about the communities in which we live—where to buy various things, where to go for entertainment, and what areas may be dangerous to walk alone in at night. A stranger entering the community would need to gain access to such information. An entrepreneur might seek to gain economically from that need by presenting information of that kind in the form of a travel guide. The author of such a guide can, under many legal systems, obtain copyright protection for his or her particular presentation of that information, but not of the informational content as such. Others are free to make use of the information contained in the guide or develop independently their own guides.

Under normal circumstances, we would expect there to be a free disclosure of such information when requested by a stranger we meet on the street. Who can be said to own such knowledge? While admittedly we would not ordinarily speak of anyone owning knowledge in this type of situation, how would the concept of ownership apply? We have already seen that a crucial test of ownership is found in the right to exclude others from access without permission. Anyone who can be said to have a right to that information can be said to be joint owner of that knowledge. In this sense, one can be said to be the owner of knowledge that one does not presently possess. The knowledge in such cases is jointly owned by all the members of the community who have a right to know.

Such community-owned knowledge is to be differentiated from the collectively owned knowledge of a society. It differs in that there is no one delegated the authority to determine who shall have access. Each member of a community must decide for himself or herself who is entitled to know. With collectively owned knowledge, agents of the government acting on behalf of the community may exclude members from access to certain kinds of information on such grounds as national security.

What values underlie the institution of property? My analysis up till now has clearly been predicated on the social utility of property rights. The social usefulness of property rights is so pervasive that it is difficult to conceive of a society that recognized no such rights. Of course, what objects to recognize as constituting property, what precise rights to recognize in relation to those objects, what limitations should be placed on those rights, as well as who should be the owner of those rights, can be subject to serious dispute. While considerations of social utility are basic in answering these questions, other values—such as justice and fairness—are also relevant.

Justice is based on giving people what they deserve, what they are entitled to, and what they have the right to expect. Someone who makes a socially valuable contribution may deserve to be rewarded as a matter of justice. The recognition by society of proprietary rights in the products of one's labor can be considered one way of fulfilling justice-based obligation. Considerations of fairness are closely related to justice. To many, it seems unfair for people to derive the benefit of the work of others without contributing in some proportionate way to the costs involved. Without such contribution the beneficiaries could be considered free riders who are unfairly enriching themselves at the expense of others.

For purposes here I will assume that justice, fairness, and utility are the basic values underlying the recognition of property rights. While they are distinct values, they function in closely interrelated ways. In many situations justice, fairness, and social utility will mandate the same result. Thus it can be argued that those who through hard work come up with socially valuable discoveries are entitled to some compensation or reward as a matter of justice, fairness, and utility. In other situations a tension may appear between considerations of justice, fairness, and utility. This is not to say that these are wholly independent values. John Stuart Mill, for one, argued that principles of justice actually form the most basic part of the requirements of utility. Without necessarily accepting Mill's position, it can be argued that the concept of social welfare or utility cannot be fully analyzed independent of notions of justice and fairness. Perceptions that one is being treated unjustly or unfairly are clearly major sources of dissatisfaction and unhappiness. If, in addition, justice and fairness are viewed as intrinsically valuable, their realization not only will be a means for achieving welfare but will form a constitutive part of it.

Since knowledge is generally valuable, we want both to encourage the discovery and development of new knowledge and to see that the distribution of such knowledge is made to all those who may benefit. We want to accomplish this while being just and fair to all those concerned. The problem is, of course, how to achieve these goals.

In the following section I will examine how the legal system has sought to deal with these problems. The final section will then explore how the norms relating to the free sharing of knowledge, which govern university-based scientific research, come in conflict with the norms of commercial enterprises seeking to exploit such knowledge for private gain.

Legal Basis

The philosophical foundation that underlies the legal recognition of intellectual property in this country comes from article 1, section 8 of the U.S. Constitution, "to promote the progress of science and useful arts." The clause clearly embodies a utilitarian justification for recognizing such rights. It does not speak of the natural or moral rights of authors and inventors to receive the benefits of their creations, but of the power of government to recognize some limited rights in order to achieve a certain goal. Underlying the system is, as the U.S. Supreme Court has stated in Mazer v. Stein 347 U.S. 201, 219 (1954), "the conviction that encouragement of individual effort by personal gain is the best way to advance public welfare through the talents of authors and inventors" (Dumbauld 1964).

The rights involved are created by the legislature as a matter of social policy rather than as recognition of some moral or natural rights held by individuals as a matter of justice. If Congress were to decide that the promotion of science and useful arts did not require recognizing such proprietary rights, it would not appear violative of any constitutional provisions (Davis this volume). That constitutional question has never arisen though, as Congress has, since the very founding of the country, considered it important to recognize such proprietary rights. Most of the colonies had patent laws, and Congress during its very first session in 1790 passed such a law. While the establishment of proprietary rights in knowledge is a matter of policy, courts in overseeing whatever legislation is passed can evaluate such laws to see that they do not violate principles of justice and fairness. Thus, if patent laws provided that only white males were entitled to receive proprietary rights in their creative work, it would clearly be violative of the equal-protection provision of the Constitution.

The belief in the value of a patent system as an incentive for encouraging new inventions and discoveries is widely shared, although not universally so. Some economists argue that patent protection is an undesirable form of monopoly. In addition, they question whether it is necessary. It has been argued that most research is now conducted by large corporations who already have sufficient incentives to make new inventions. Being first with a new product by itself creates a substantial advantage over competitors.

The granting of exclusive rights for a limited time can be seen as a compromise between giving the creator of new knowledge full and unlimited ownership rights and giving the public free access. One way of looking upon the patent law is that it is a kind of contract between the government and an individual in which, in exchange for public disclosure of one's invention, a person is given exclusive rights to make use of it for a limited period of time. Under common-law doctrine, which still plays a vital role, a company's trade secrets (that is, knowledge it uniquely possesses that gives it an advantage over competitors) are legally recognized as property. Such trade secrets can include patentable inventions as well as other forms of knowledge, such as customer lists, new product plans, and cost and pricing data. The company is protected against wrongful disclosure to competitors by former or present employees. Trade secret protection however does not prevent a competitor from independently discovering such knowledge and even obtaining, where applicable, a patent for exclusive rights to use that knowledge. A company sometimes has to decide whether to seek the limited protection provided by patent law or to rely on the protection of trade secrecy, which in principle is unlimited in time. The law, while wanting to encourage disclosure, does not mandate it. The fundamental right of nondisclosure is respected and that right can be said to underlie the law of intellectual property.

The right of nondisclosure is somewhat difficult to characterize. That limited right is not sufficient to make one the sole owner of the knowledge one possesses, as others are entitled to gain access independently to it as well. As one does not have a duty to disclose what one knows to others, it does not constitute jointly or collectively owned knowledge either. The basic right of nondisclosure seems more like a right of privacy or autonomy than what we ordinarily consider a property right. It is a right of possession rather than of ownership because others have the right to gain possession as well without obtaining permission. At issue is what sorts of knowledge should have further exclusive ownership rights and under what conditions.

The patent laws do not protect basic scientific discoveries but only the method for making practical use of them. Roughly speaking the patent law protects knowledge only when it constitutes recipes for making things. This limitation in the patent law has profound effects. Can it be justified? One can argue that in terms of the basic values of justice, fairness, and social utility this restriction is not justified.

While at an earlier time most advances were derived from craft and guild know-how, today theoretical research is a major source of technological developments. Thus the new biotechnology clearly arose out of basic theoretical discoveries and would not have occurred without them. In terms of merit or entitlement there is no reason to believe that those who made the basic discoveries are not at least as deserving of the fruits of their labor as those who made the technological developments. Without such compensation, those responsible for the technological development can be considered free riders who benefit from the labors of those who undertook the more basic research. This seems fundamentally unfair.

In terms of social utility, it can be argued that the present patent system fails to provide adequate incentives for doing basic research, which usually bears only an indirect and uncertain relation to practically useful products. In the long run, however, such research is highly beneficial to society. The economist Frank Knight has argued that the patent system misdirects rewards for innovation by rewarding the "routinizer," who usually only takes the last step in the creative process, and not those doing the pioneering work (Knight 1957; Bowman 1973). This judgment may be somewhat overdrawn, as technological discoveries can involve a great deal of their own kind of creativity. Still the patent system provides little incentive to undertake basic research as compared with technologically cultivating existing theoretical knowledge.

While there may be some reasons of principle for not recognizing proprietary rights in theoretical knowledge, one may suspect that the main arguments against such recognition relate to the complex administrative problems that would arise. There would be enormous problems deciding what kinds of theoretical knowledge were patentable, resolving conflicts over claims of priority in discovery, determining questions of infringement and methods of compensation. Many inventions are based on a large number of theoretical discoveries; this further complicates the problem. The administrative costs of the present patent system are already quite substantial and would be greatly increased if the coverage of the system were expanded in this manner.

One way of avoiding some of these administrative problems would be to impose a tax, based on a percentage of the profits obtained by patented inventions, and with the proceeds establish a fund. The main purpose of the fund would be to support basic research, though it could also serve to reward those who have made fundamental contributions.

One consequence of the nonpatentability of theoretical knowledge is that most commercial corporations do not engage in basic research. With a few notable exceptions—such as the Bell labs—corporations consider it too risky an investment. They are uncertain whether their researchers will come up with knowledge that can turn out to be commercially useful. There is also the fear that others can freely appropriate that knowledge in obtaining their own patents.

If other institutions in our society (specifically universities) were not committed to the furtherance of basic knowledge, the problem of the undersupport of basic research would be considerably greater. Even within universities, much of the research undertaken is an adjunct to the function of transmitting knowledge and supervising graduate students seeking advanced degrees. Up until World War II, scientific research was essentially subsidized by universities themselves; since then the government has been a major source of funds. Because the distinction between basic and applied scientific research is sometimes unclear, much basic research has been done under grants nominally classified as applied research. The Reagan administration, as part of a general policy of cutting back on government programs, sought to reduce such support. Universities, anticipating this decline of government support, began looking for alternative sources of funding. It was about this time that the potential of the new biotechnology became the subject of a great deal of publicity, and venture capitalists were exploring ways of exploiting its commercial possibilities.

The University Context

Universities and commercial enterprises are clearly different kinds of institutions, having distinctive goals and norms. The problem is one of regulating their interaction so as not to affect adversely the institutional integrity of either.

If anything is considered fundamental to the ethics of scientific research as conducted within a university community, it is that there be a full and free sharing of knowledge. Scientific progress is based on a collaborative process in which many individuals make contributions of varying degrees of significance. Until certain hypotheses are critically scrutinized by others within the relevant community, there is little ground for accepting these beliefs as authentic knowledge. It is true that at certain stages of their work scientists may be secretive in order to ensure that they will receive the recognition and honors given those who make original contributions. Still, they are clearly committed to public disclosure as soon as they are satisfied with the quality of their work.

Commercial enterprises, on the other hand, are not committed, as a matter of principle, to the free sharing of their knowledge. While they may share certain kinds of information through trade associations as a matter of policy, it is an accepted part of the norms of fair competition for companies to take advantage of their trade secrets, which can be among the most important assets a company possesses. As I have already indicated, a commercial enterprise often has a choice between obtaining a patent on its inventions or relying on trade secrecy protection. Even when a company applies for a patent, it seeks to disclose as little as necessary in its claim. Finally, a company that does receive a patent may decide neither to use nor to license others in the use of that new knowledge when it is judged in the economic interest of the company. Companies have even been known to purchase the patents of others for the very purpose of preventing their introduction into the market.

The fundamental theoretical discoveries made in the 1950s concerning the genetic structure of DNA were followed in the 1970s by the development of recombinant DNA techniques that make it possible to genetically engineer new microorganisms. This new biotechnology is not only potentially beneficial to society, but is expected to be a source of great economic wealth. The Supreme Court has held that the products of such bioengineering are patentable; this further increases their economic value.

Much of the work in this area has been done by academically based molecular biologists. Some leading scientists have left academic institutions to carry out their research and development activities in newly created commercial enterprises—such as Genentech, Biogen, Cetus, and Genex—or with established pharmaceutical manufacturers. In other cases, special ties have developed between universities and commerical corporations—such as Harvard and Washington University of St. Louis with Monsanto, Massachusetts General Hospital with Hoechst, and MIT with an institute established by industrialist Edwin Whitehead. In exchange for providing substantial financial support, the companies receive a variety of things, ranging from patent rights or exclusive licenses to (in the case of the Whitehead Institute) participation in faculty appointments and control of research areas. In addition, they may exercise varying degrees of control over publication of results (Wade 1984).

Aside from these special arrangements, a large number of university-based molecular biologists have developed connections with commercial enterprises as consultants, working during the time they are considered free from their university responsibilities. Such interaction between academically based scientists and commercial enterprises is not new. What is perhaps unusual is the extent of the involvement, the rapidity with which it has evolved, and the potential economic benefits. Basic scientific research does not generally lead in such an immediate and direct way to large-scale commercialization. This has raised in a dramatic way some basic policy issues concerning university-industry relations in general and specifically whether certain values underlying the conduct of basic scientific research are being put in jeopardy.

One of the basic issues raised by joint ventures between universities and businesses is control over publication. It seems an almost necessary concomitant that some control will be exercised. The companies will want to have at least a preview in order to have their patent attorneys consider the desirability of disclosing certain details. They might want to delay publication until a patent application has been made, although this is not essential. (In the United States, there is a grace period of one year in which to file a patent claim after public disclosure. This differs, however, from most European countries, where one must file a claim before any publication takes place.)

While some short delay in publication may be acceptable, anything further may constitute a threat to the norms governing scientific research. The idea of a corporation exercising veto power over the publication of some scientific result would constitute an affront to established norms. What is more likely are requests by the business involved to withhold some results. This also jeopardizes the integrity of scientific research. Perhaps the greatest danger is that scientists will over time come to appreciate the proprietary point of view and will consider these kinds of restrictions justified. Scientists employed by commercial enterprises have learned to live with and perhaps accept as justified various kinds of restrictions of this nature.

It is interesting to note that some significant changes have already been reported in scientists' attitudes toward patents and free communication of knowledge. When Stanley Cohen and Herbert Boyer of Stanford University invented the gene-splicing technique in 1973, they had no thought of patenting it, and only through the urging of the university's patent counsel did they agree to do so. They consented only on the basis that the university become the exclusive beneficiary. Cohen later commented, "My initial reaction . . . was to question whether basic research of this type could or should be patented and to point out that our work had been dependent on a number of earlier discoveries by others" (Wade 1984, p. 31). Cesar Milstein, who shared the Nobel Prize for his contribution to the development of monoclonal antibody technology, or hybridoma, did inquire as to whether the method should be patented. The policy of his sponsor, the British Medical Research Council, was to make such new methods freely available to others. He later was reported as saying, "We were too green and inexperienced on the matter of patents . . . We were mainly concerned with the scientific aspects and not giving particular thought to the commercial applications" (Wade 1984).

A number of incidents have been reported in which scientists have become less free in sharing their ideas. A committee at the University of California at San Francisco reported that so many of their scientists were also employed by Genentech that "people were loath to ask questions and give suggestions in seminars or across the bench, for there was a feeling that someone might take an idea and patent it, or that an individual's idea might be taken to make money for someone else" (Wade 1984, p. 37).

In another case, two research scientists at UCLA, Golde and Koeffler, gave Gallo, a researcher at the National Cancer Institute, a valuable sample of cancerous cells they had succeeded in growing. It was then passed on by Gallo to a scientist friend working for a biotechnology company who found a way to make use of those cells in the production of interferon. A dispute later arose as to whether permission had been given for that further dissemination of the cells. After the incident, Golde is reported as saying, "Everything has changed. The exchange of materials is different. They now have value. To send out a cell line for some experiment is like sending out a 20-carat diamond to cut some glass" (Wade 1984, p. 36). Nicholas Wade states, "Gallo, whose policy of making cells freely available to other researchers was well known, now says he will send nothing out unless he owns it entirely or has written permission from everyone involved" (Wade 1984). A law suit by Golde and the University of California against Hoffmann–La Roche and Genentech, which was finally settled out of court, arose out of this incident.

Concern over patentability can also effect decisions as to areas of research. Leading scientists may shift their areas of research, as well as that of their graduate students, based not on considerations of theoretical importance for the advancement of knowledge but of economic profitability. If a patent on a certain method and product of genetic engineering has already been obtained, other researchers may become hesitant to do further research in that specific area. There could be problems concerning patent infringement. One would have to study carefully the precise claims contained in the patent in order to see whether one can design around them. Scientific progress often depends on a continuous process involving slight refinements and improvements. The patent law may prevent this from taking place. Even if a scientist came up with a significant improvement that is itself patentable, the chances are that he or she would have to enter into a licensing arrangement with the holder of the basic patent to make use of it.

Patent law is highly complex, and there is a great deal of discretion exercised by the Patent Office and the courts in its administration. At the beginning stages of a new technology, the tendency is to permit broad claims. As the technology develops, claims are more narrowly defined. Thus the Cohen-Boyer patent is quite broad, but there are indications that the Patent Office is taking a more restrictive view with hybridomas. It is also interesting to note that Cohen and Boyer were not able to obtain patents in most European countries because they published their work before filing their patent claim (Office of Technology Assessment 1984, chap. 16).

In addition to having a patentable subject matter, one must establish that the invention is new, useful, and nonobvious. An invention that would have been obvious to someone of ordinary competence in the field is not patentable. If inventions that are not genuinely new and nonobvious could be patented, the public would in effect be deprived of the free use of what it already knew or what was an obvious extension of what it knew. Because of the rapid developments taking place in biotechnology, it may be difficult to say at any particular time what is new and nonobvious. A considerable number of contested patent claims and court suits are likely to arise.

If scientists within academic institutions take the intricate requirements of the patent law into account in structuring their research projects, there will be a significant departure from established practice. The proprietary view of knowledge could have a very damaging effect both on the way scientists share information and on how they make decisions as to what areas to explore. While the actual practices of scientists never fully conformed to the normative ideals, the potentially great economic rewards involved here add a new and possibly destructive dimension.

How serious are these dangers? The dramatic examples cited constitute anecdotal information, and there is some danger of overgeneralizing from a few cases. A survey was done of microbiologists at six leading research universities and fifteen biotechnology firms for the Office of Technology Assessment of the U.S. government. Two of the questions asked were, "Has the way university research is done or the quality of university research been affected by the relationships?" and "Has collaboration among university researchers been affected by the relationships?" About 85 percent of the respondents maintained that neither the quality of the research being done nor collaboration among university researchers has been adversely affected by these recent developments (Office of Technology Assessment 1984, pp. 413ff.). It is not fully clear what inferences one should draw from that survey. Does the fact that 15 percent felt it was adversely affected have any significance? There is clearly a need to explore in a more systematic way what is taking place before one decides what can and should be done.

Assuming that such changes are taking place, are they necessarily undesirable? Will it distort judgments about research areas and hence retard further basic scientific discoveries? While it may be reasonable to assume that it would have some adverse effect, one cannot be certain. The increase in competitiveness among scientists and universities resulting from proprietary rights in knowledge could conceivably result in greater knowledge, both basic and applied.

Concern over changes in the way scientists conduct their research, however, is not based solely on consideration of possible undesirable consequences. Those changes put into question certain basic institutional values identified with a university and forming part of the way of life of scientists working in such communities.

This was revealed in a dramatic way in 1980. Harvard University administrators conceived a plan for establishing a biotechnology company in partnership with its own professors. A memorandum in support of the proposal cited a number of benefits. One was the substantial financial gain to the university; another, the fact that microbiologists were in any case going to get involved with industry, and the university could be an effective advisor and negotiator for the faculty; and third, such a company could avoid the drawbacks of commercial companies with their "excessive secrecy requirements, undue direction of the work of scientists, and total separation from the atmosphere of a research university" (Wade 1984, p. 42). One of the drawbacks cited concerned the possibility that such a venture might encourage faculty members to shift their areas of interest to what was commercially marketable. Another major drawback related to the feelings of resentment that would arise among those faculty members in microbiology who were not to be included in the enterprise.

Ten members of the Cellular and Developmental Biology Laboratory objected to such an arrangement. It would, as one said, "deflect the University from its central and essential function: to advance learning, to foster free and searching inquiry into fundamental problems, and to communicate that learning, and the spirit of free inquiry, to the oncoming generation of students" (Wade 1984, p. 44). This is a succinct statement of the institutional values that are at stake.

A different kind of objection was raised by Walter Gilbert, a Nobel Prize winner who left Harvard to become president of Biogen, a biotechnology firm. He essentially complained that it would constitute unfair competition. "I have my own company and I would resent being put in the position of having to compete against Harvard for the best people and best work. I might have to push Harvard to the wall in some cases of competition." He also pointed out that the logical extension of the university's involvement would be "to convert other areas of the university over to profit-making ventures, start a law firm in the law school, have the English department write advertising copy, develop a way for the doctors to earn extra profits at the medical school . . . The idea is completely mad at a certain level" (Wade 1984). The proposal was dropped within a month after being presented to the faculty.

How considerations of fairness relate to these kinds of situations is complex and depends in part on one's point of view. Is it fair to a university if a scientist, who is being provided with research resources, chooses to leave when that work becomes commercially valuable? Scientific researchers working for commercial enterprises are usually regulated, by contractual agreement, as to what use they can make of the results of their research once they leave. Even without contractual agreement, a court might on equitable grounds prevent a scientist from making unfair use of the knowledge obtained while employed. The law has, however, to balance notions of fairness against the importance of not unduly restricting an individual's freedom to make changes in employment.

Would a university be protected under the laws of trade secrecy? That protection is generally limited to profit-making businesses. It has been reported, however, that five states have developed a broader concept of trade secrets under which the results of basic research carried out within a university might be protected. Whether that is a desirable development is difficult to say. The time has perhaps come when universities will have to establish policies regulating the exploitation for personal gain of knowledge obtained while doing scientific research as a member of a university community.

Much of the basic research done within universities in molecular biology has been funded by the federal government. Up until 1980 the federal government was entitled to receive the rights to all patentable inventions arising out of government-sponsored research. In 1980 an amendment to the General Patent Law was passed that gave universities, as nonprofit institutions, the right to hold such patents. The government retains for itself only the right to use those inventions without paying any license fees. The purpose of the change was to provide more of an incentive to develop and commercially exploit the inventions that resulted from government support. Most large universities now have their own patent counsel and arrangements with companies that specialize in commercially exploiting newly patented inventions.

While the main function of universities is to advance human knowledge, they do compete with one another. The competition among universities over obtaining valuable patent rights is an additional factor that may affect the way colleagues located at different universities cooperate with one another. This competition is not only on the institutional level; it is common practice for a university to give the researchers who made the patentable discovery a certain percentage of the proceeds obtained through licensing agreements.

This raises the question of what use should be made of the money received under such licensing agreements? Perhaps it should be said that it may not turn out to be so much as some people expect. While royalties from the Cohen-Boyer patent have already produced $2 million, later and more narrowly defined patents are likely to produce less.

Assuming that a university does receive substantial royalty fees, what should be done with the money? The basic values of justice, fairness, and social utility again come into play. The inventor deserves as a matter of justice some reward for the contribution that benefits the university. Some proceeds would also be given to the discipline or area in which the work was carried on. Inventing is now largely a social activity, and colleagues have probably—directly or indirectly—made some contribution to the discovery. This money would also serve as an incentive for doing further research in that area. As a matter of fairness, it might be appropriate for a portion of the royalties to be given to colleagues in related disciplines whose basic, nonpatentable knowledge was essential to the discovery. Is there any responsibility to distribute a portion of those funds to further research in disciplines that are completely unrelated to the scientific enterprise? How that question is answered reveals in some small way a university's conception of itself as a community. Values of community are perhaps distinct and not reducible to the values of justice, fairness, and social utility.

Acknowledgment

The author gratefully acknowledges support received for the research involved in this chapter from the EXXON Education Foundation through the Center for the Study of Values and Social Policy of the University of Colorado.

References

Bowman, W. S., Jr. 1973. *Patent and Antitrust Law*. Chicago: University of Chicago Press, pp. 28–31.

Dumbauld, E. 1964. *The Constitution of the United States*. Norman: University of Oklahoma Press.

Knight, F. 1957. *Risk, Uncertainty and Profit*. Boston: Houghton Mifflin.

Office of Technology Assessment. 1984. *Commercial Biotechnology*. Washington: Government Printing Office.

Wade, N. 1984. "Background Paper in *The Science Business Report* of the Twentieth-century Task Force on the Commercialization of Scientific Research." New York: Priority Press.

15. The Intellectual Property Literature: A Structured Approach

Patrick Croskery

The specifically ethical study of intellectual property is a newly emerging topic, and as such does not have a well-established literature. There is, however, a large body of work by lawyers, economists, philosophers, and others that discusses important intellectual property issues in a normative fashion. The purpose of the annotated bibliography that follows this essay is to give the interested reader access to these works.

One encounters two difficulties in bringing this diverse literature together. First, these writings are wedded to various large contexts. For example, economists model the patent system as part of the larger goal of understanding the economics of innovation and invention. Lawyers, on the other hand, are writing in the context of careful, detailed consideration of existing law. Normative claims arise in the course of this work, but usually do not form the greatest part of a legal analysis. Sociologists considering the effect of the patent system on university research are often doing so in the context of the study of the sociology of scientific research in general. Philosophers generally approach the topic of intellectual property from the perspective of property theory.

For the reader who becomes interested in one of these approaches to intellectual property, the larger context will be quite important, However, to include the relevant literature for all of these contexts would make the bibligraphy hopelessly unwieldy. My solution to this problem has been to select the articles that are most capable of standing on their own, to focus on those which make relatively clear-cut normative claims, and to restrict the bibliography to a limited number of topics, so that each can be covered to a reasonable degree without making the bibliography difficult to browse through. For example, the interaction between patent and antitrust law is not covered in this bibliography, despite its obvious importance. The bibliography's introductory note lists the topics it includes.

The second problem is a consequence of the first. Since they arise in such different contexts, it is not clear how the various accounts of intellectual property relate to one another. In this chapter, I offer a structured way of categorizing these accounts that allows them to be compared in ethical terms.

An Overview

One assumption I will be making throughout this chapter is that there is a class of goods that can be perperly called *intellectual property*. I will call this hypothetical class of goods *IP goods*. There are several points that are relevant while reading accounts of the ethical issues related to these goods. First, as I have suggested, it is important to keep in mind the context in which the goods are being viewed. Second, we need to be aware of the features that IP goods are taken to possess. In this paper I will suggest that these two points are interdependent; that is, that in different contexts, IP goods are seen as possessing different characteristics. The debates, then, are as much about what IP goods are as about how they should be treated.

The term *good*, here, is taken from economic discourse, and refers simply to some specifiable entity that contributes positively to welfare. At least initially, ordinary intuitions about what constitutes a good (for example, a car or a house) should be sufficient, providing that this set includes intangible goods (such as a text or an invention).

There is, however, a risk that using the term *good* will create more confusion than clarification. Once we move beyond dealing with everyday material objects, there is much room for interpretation. For example, is each copy of the text of a book a separate good, or is the text, which has many inscriptions, just one good? Our ordinary intuitions about goods do not apply cleanly in this case. It is important when thinking about intellectual property goods to keep this sort of difficulty in mind. For an alternative way of thinking about the relation between intellectual property and more familiar goods, see Davis (1987).

A Closer Look at IP Goods

There are two aspects of IP goods that are particularly relevant to accounts given of these goods. First, it is often suggested that they have the feature of *nonrival consumption*. If a good is nonrivally consumed, then one person's use of the good does not diminish the use of that same good by another person. For example, two people can be reading the same book (or rather, two inscriptions of the same text) without interfering with one another. Of course, two people can be riding the same train without interfering with one another as well. But two people cannot be riding in the same seat on the same train without interfering with one another, and we cannot simply make an extra copy of the train seat for the second passenger.

There is another way of expressing this distinction. We might say that the underlying good in the first case—the text—is not consumed in use, whereas the underlying good in the second case—a seat traveling from *a* to *b* at time *t*—is consumed in use. A text might be consumed in another sense, however; all of its profitable uses might be exploited. A large printing of a book might completely fill the demand for that book, so that there is no further value in the good. When thinking about the consumption of IP goods, it is important to look beyond the mere fact that they can be copied.

A second useful distinction is between goods which are created and those which are discovered. Which feature characterizes IP goods is important to the ethical intuitions that bear on those goods. If someone merely discovers a good, we do not feel that he has so strong a claim to it as someone has to a good that she has created. Both sorts of claims have historically justified property rights, but creation seems intuitively to be a stronger justification. Of course, by choosing difficult cases, where a great deal of labor is invested in discovering a good and a small amount of labor is invested in creating a good, we can diminish the clarity of our intuitions. However, taking relatively straightforward cases, the discovery of a good does not seem to create so strong a right as the creation of a good, and deciding whether an IP good is created or discovered will affect our treatment of it.

Utilizing these two distinctions, we naturally get four possible characterizations of IP goods. For each of the characterizations there are goods other than IP goods that uncontroversially meet the criteria. These other goods are handled by specific mechanisms that affect their production (or discovery) and distribution. This set of relationships is most easily seen in a figure; for simplicity of reference, I am assigning each category of good a name suggestive of its properties.

An ordinary good is something, like a match, that is created and rival. A car also fits in this category; that is, consumption does not have to occur at once. These goods are exchanged by their creators. This exchange process can be quite complex, as in a modern capitalistic economy, but pointing out the simple

Figure 15.1. Categories of Goods

	Rival (consumed)	Nonrival (not consumed)
Created	Ordinary goods: match car [mechanism = exchange]	Copyable goods: rules of a game business procedure [mechanism = award] (proposed)
Discovered	Natural resources: coal fish [mechanism = prospect]	Costless goods: law of nature mathematical truth [mechanism = universities]

mechanism of exchange is sufficient for my purposes. There are complications when this mechanism is applied to IP goods, and many discussions of ethical issues related to intellectual property turn on these complications.

Costless goods are those which have not been created, and are nonrival. I call them *costless* because they have no costs of production and distribution. An example, admittedly a bit difficult to get a hold of conceptually, is a law of nature. For this example to work, the law of nature must be understood as the actual regularity in nature, rather than, say, its formulation in the notation of a specific scientific theory. People do not create such laws of nature, yet they make use of them. Clearly, one person's use of a natural law does not preclude any other person's simultaneous use of that law.

Laws of nature and other costless goods are usually treated by the research university system, a complex network composed of private and public universities, supported in part by governmental grants. While calling this network a *mechanism* is somewhat misleading, it does seem to be as coherent a system in many ways as an exchange economy. Its dynamics are quite different, however. Within the context of the university system, a researcher does not directly pay another researcher for the use of his ideas. Rather, he officially recognizes her contribution, and she is rewarded in the form of increased prestige. This prestige can translate into additional salary, further grants, or simply be desirable for its own sake. Ethical issues concerning intellectual property arise when researchers consider using the patent system to receive more direct rewards for their efforts.

Natural resources are those goods which are rival but are not created, such as coal. The mechanism that normally assigns rights to these goods, the *prospect* mechanism, assigns the rights to the one who demonstrates, according to some formal procedure, the prospect that they will be found. When a gold miner stakes his claim, for example, he needs to demonstrate that gold exists in that claim, by submitting a sample. Once the rights to the resources have been assigned, and the goods have actually been acquired, those resources can be exchanged just like ordinary goods. The addition of the prospect mechanism is made necessary by the fact that in the case of natural resources there is no creator to initially own the good.

The relevance of this mechanism to my discussion is largely due to the arguments of Edmund Kitch, who suggested in an influential 1977 article that the patent system can be understood as a prospect mechanism. This claim has two elements: first, the descriptive claim that this is the way the intellectual property system *is;* and second, the normative claim that this is the way that the intellectual property system *ought to be.* Kitch makes both claims, but my discussion of his views later in this chapter will be concerned with only the second, normative claim.

The last category of goods—copyable goods—are those which are created but nonrival. The fact that these goods can be copied is what allows them to be nonrival. A firm can copy another's business practices, such as offering rebates. Such copying is considered to be an important part of a competitive marketplace. There is no mechanism that rewards the production of copyable goods of this sort, although lead-time advantages can make it worthwhile to produce them. (That is, a firm can obtain profits from an innovation while other companies are in the process of copying that innovation.) Some people feel that IP goods also fit into this category, and suggest that an entirely new mechanism be developed for them that reflects their status as copyable goods. Alternatively, this line of reasoning could be used to suggest that IP goods receive no protection, for the same reasons that apply to other copyable goods.

In the rest of this chapter, I will spell out in somewhat greater detail how the classification of goods and mechanisms shown in Figure 15.1 applies to the more specific ethical issues considered in this volume and those found in the bibliography. To keep my discussion to a reasonable length, I will have to give quite succinct accounts of points of view that are very complex. I will draw out the features relevant to my overall framework, and will not attempt here to consider all the potential objections and responses to these perspectives and my discussion of them.

IP Goods as Ordinary Goods

Such ordinary goods as cars or houses are owned by their creators-producers and exchanged for one another. The use of money instead of barter complicates the picture, and ownership is itself often quite complex in a modern legal system. In general terms, however, only two facts need be singled out: that the creator owns the good, and that the good can be exchanged. IP goods, when seen as created and rival, would seem to deserve the same treatment as other ordinary goods. Thus, one set of perspectives on these goods argues for such treatment.

A great variety of positions have been taken within this rough frame. At one extreme on the issue of ownership is the doctrine of *droit moral*, an important concept in French copyright law (although not in U.S. law). This doctrine holds that the creator of a work has certain inalienable rights, such as the right of publication. Since these rights are inalienable, they are in fact stronger than those associated with such ordinary goods as cars—they cannot be taken away from the creator.

In U.S. law, in contrast, the rights given by the copyright and patent systems are in some respects weaker than those associated with ordinary goods. The most common account of these rights in U.S. law is that they represent part of a bargain between society and the author or inventor. According to this account, for example, an inventor is granted a monopoly on the use of his invention for a period of seventeen years in exchange for disclsoure of his invention and the surrendering of it to the public domain after that period. By treating the creator's rights as conditional in this way, the bargain account implicitly supposes that they are weaker than those rights associated with such ordinary goods as cars, which are not conditional on a bargain. At its most extreme, the bargain account need not acknowledge that an author or inventor have creator's rights at all. However, in most accounts of U.S. law, there is concern for the rights of the creator, even where those rights are somewhat weaker than ordinary property rights. A notable exception is Edmund Kitch's theory, which is, as a consequence, considered in a later section of this chapter.

Between these two extreme positions lie many more, which pay varying degrees of attention to the creator's rights in his creation. The common element of these accounts is that, because they see IP goods as created, they must deal with the issue of the rights of the creator. One right that is of particular significance is the right to profit from a creation via the marketplace.

It is at this point that the nature of IP goods creates a few special problems. If the creator sells a copy of his creation to someone else, he cannot be sure exactly what he has sold. It might be that he has sold one copy. On the other hand, if the buyer makes copies and distributes them, then the creator has (in effect) sold the buyer many copies. The rights given by intellectual property law resolve this ambiguity, by preventing the buyer from making additional copies unless he makes arrangements with the creator (or someone to whom the creator has sold or assigned the rights).

Exactly how to characterize the rights involved is a delicate question. The most common characterization is that the creator is given monopoly rights. However, it is a monopoly that one always has in the case of ordinary goods: I have the same monopoly on my car, since no one is able to make copies of it. Cars with the same design can be made, but that is copying the intellectual property aspect of the car. In the case of my car, the restriction is placed by nature, whereas in the case of an IP good, the restriction is placed by law.

There is another perspective available on this issue, which can be understood by considering again the complications of talking about the good that is an IP good. One has to distinguish the selling of a copy (or inscription, or instantiation) of an IP good from the selling of the good itself. One way of looking at intellectual property law in this context is that it gives the creator of the good the right to sell instantiations of his good without selling the good itself.

Using this contrast between instantiations of an IP good and the IP good itself, it is possible to argue that the IP good is consumed in use, just as ordinary goods are. This can be seen most clearly in a case where the potential uses of an

IP good are quite limited. An invention, for example, might be valuable only to manufacturers of lawn mowers, and only while particular methods of manufacture are in use. Once it has been applied in those cases, it has been, in a sense, consumed. It is no longer of value.

In principle, this argument can be extended to cover all IP goods, since they all have a finite number of uses. If an IP good is going to be of value into the indefinite future, however, the argument seems to fail. This sort of difficulty is a common one in economic thought, and the standard response is to discount the future value when calculating present value. Using such a mechanism, it is possible, in theory at least, to determine the present value of an IP good. It is in relation to this value that we make decisions about how much we should spend in creating the IP good, and so it is presumably that value which represents the point at which the good is consumed.

The idea that IP goods might in a sense be consumed in use is a complex one, and I will consider it again briefly when I look at the category of natural resources. There are a number of other issues that could be considered under the general heading of IP goods understood as ordinary goods, but the discussion up to this point should be enough to give a sense of the ethical implications of this perspective. For the reader who wants to explore accounts that treat IP goods in roughly this way, I would recommend the following articles from the bibliography:

(1) On the theory of the moral right of creators, *droit moral:* DaSilva; Dworkin; Kwall; Roeder; Strauss; Valentin.

(2) The work of economists modeling and otherwise examining the patent and copyright systems: Arrow; Beck; Dasgupta, Gilbert and Stiglitz; DeBrock; Fellner; Fudenberg, Stiglitz and Tirole; Gallins and Winter; Harris and Vickers; Horstmann, McDonald and Slivinski; H. G. Johnson; Judd; Kamien and Schwartz; Katz; Levin; Mansfield; Mansfield, Schwartz and Wagner; Mansfield et al.; Nordhaus; Novos and Waldman; Reinganum; Scherer; Shapiro; Tandon; Wright.

(3) Historical discussions: Batzel; Crawford; Dutton; Machlup and Penrose; Mark; Patterson; Sherwood.

(4) Expositions of property theory (in general): Becker (1977, 1987); Davis; Honore; Pennock and Chapman.

(5) Discussions of the status of new technologies, such as software: Bender; Canfield; Chesser; Chisum (1986); Davidson (1983); Dunn; Goodman; *Harvard Law Review* (1983); D. G. Johnson; Keplinger; Kidwell; Lahore, Dworkin and Symth; Mislow; National Commission on New Technological Uses of Copyrighted Works (CONTU); Raskind; Samuelson (1984, 1985); Schmid; Webster.

(6) General discussions: Branscomb (1985, 1986); Brown; Esezobor; *Harvard*

Law Review (1982); Machlup; Stedman; Office of Technology Assessment; Wincor; Woodward.

(7) Miscellaneous discussions: Beier, Beier, Crespi, and Straus; Dratler; Kayton; Kozinski; Lehmann; Lunn; Massel; Nimmer (1975); Nordemann; O'Hare; Oppenlander; Reich; von Hippel.

(8) On trade secrecy (a topic that I did not discuss, but that seems to fit in this category): Bender; Bok (1982b); Cheung; Dorvee; Frederick and Snoeyenbos; French; Kitch (1980b); Klitzke; Lieberstein; Milgrim; Soltysinki; Stevenson.

(9) Reference works: Bush; Chisum (1978, 1980); Elias; Kitch (1986); Miller and Davis; Nimmer (1978, 1985); Roberts.

(10) In this volume: Boonin; Cranor; Davidson; Davis; Dreyfuss; Dworkin; Kelly; Kuflik; Samuelson; Winter.

IP Goods as Costless Goods (Laws of Nature)

Calling laws of nature *goods*, sounds a bit peculiar, but characterizing them as nonrival and discovered should make sense. Since it does require effort to discover these laws, and their discovery often provides significant benefits to society, it is not surprising that a mechanism for discovering them in an organized fashion has evolved. Whether one wants to think of research universities as a "mechanism" for handling these "goods" in the same way as the marketplace handles ordinary goods is an issue that I will not go into in this paper. In any case, it should be noted that the following characterization of the university research system is a very simplified and idealized one.

Costless goods are primarily handled by the university-government research system. This complex system uses prestige (or honor) as an incentive for discovery, and also as a means for assigning grants, with the goal of maximizing productivity. A researcher who establishes a good reputation will receive increased support for her research, and a higher salary as well, on the implicit assumption that she is likely to perform valuable research in the future. In addition, these rewards motivate other researchers, who hope to receive such rewards should their work prove to be valuable. In this way, discovery is rewarded, although somewhat indirectly: there is no explicit buying and selling of discoveries. This system could be discussed metaphorically as a system of ownership, but the differences are important to keep in mind. In particular, use is free in the research system, and this precludes an exchange based assessment of the value of a good.

That use is free reflects the fact that the goods involved are nonrival. When considering rival goods, an important question is distributional: Who should be allowed to consume the good? In the case of nonrival goods, there is no reason to exclude anyone from consuming the good, since that use will not diminish the use of the good by anyone else. (This position is quite contrary to the view considered earlier in this discussion, which holds that there is an important sense in which apparently nonrival goods are consumed in use.)

The university research mechanism involves more complex incentives than the marketplace (at least at first glance); noneconomic motives play an important role in university research. However, it is also clear that the person who discovers a natural law is not able to capture all the economic benefits that flow from that discovery. This reflects, in part, the fact that discovering a good does not create so strong a right to that good as creating it. Of course, it also reflects the fact that we do not think of natural laws as goods, as well as the view that granting rights to entities as broad as laws of nature would hinder rather than encourage scientific development.

In a general way, then, one can see that the university research system reflects the two properties of the type of good it handles—costless goods. Although the university research system is not intended to handle IP goods, it often happens that IP goods become entangled in it, partly because the distinction between laws of nature and IP goods is difficult to draw. A law of nature cannot be patented, but a concrete application of that law can be. This application need not be a physical machine—processes are patentable—yet it must not cross an invisible boundary that separates an abstract idea from a concrete application. In such areas as biotechnology, the move from the discovery of a scientific principle to a practical application is often small enough that the researcher who makes the discovery is in an excellent position to patent the commercial application.

Should the researcher go ahead and patent the commercial application? In many ways, the question is a practical one: What is the effect on research of allowing or not allowing researchers to patent where possible? The question has a normative dimension primarily because the university research system relies so heavily on noneconomic incentives, and it is felt that the intrusion of a profit motive into the system undermines the norms that support those noneconomic incentives. That is, the issue represents a conflict of norms.

The norm that seems the most threatened is that of openness. In part this is because the university research system and the patent system have different methods for publicizing results. The patent system often involves much longer periods of secrecy than the university research system, and also has a more rigid procedure for determining priority. The university research system (in which reputation measures the value of someone's contribution) can recognize that discoveries are often a complex blend of the work of many individuals.

The threat to openness also results in no small part from the much greater financial reward possible through the patent system. From the vantage point of the researcher, it must seem peculiar that the small move from theory to application mentioned earlier is judged to be of such enormous value by our economic system. Another way of looking at this is to suppose that the university research system dramatically underrewards participants, and this fact is brought home to researchers when the patent system directly impinges on their work. When a researcher is contrasting the rewards of working in private industry versus working in a university setting, he can suppose that his lower payoff is balanced by the intangible benefits of working at a university. There is no equivalent balance when he compares himself to another university researcher who has obtained a valuable patent.

If one supposes that laws of nature and IP goods are not in fact of morally different sorts, then the reward differential seems inexplicable. Although I have not encountered an argument to the effect that IP goods are costless goods (that is, discovered and nonrival), it is a possible perspective, which could be justified by selectively combining arguments considered in the other sections of this chapter. Such a perspective might lead one to be critical of the differing payoffs of the university research mechanism and the patent system on normative (rather than merely pragmatic) grounds.

This account of the patent/research relationship is greatly oversimplified. To see how patents and the research system interact in more realistic cases, the following works should be consulted:

Adler; Bok (1982a, b); Carpenter, Cooper and Narin; Evenson; Fusfeld and Peters; Grobstein; Hull; Kenny; McMullin; Melman; Nelkin; Panem; Rosenzweig; Weiner.

In this volume: Buttel and Belsky; Boonin; Goldman; Lemin; Weiner.

IP Goods as Natural Resources

Natural resources are, like ordinary goods, usually exchanged in a market context. But since they are not created, they cannot be owned by their creator. Instead they are owned by their discoverer, which requires some method for defining discovery. Exactly how one defines the discoverer turns out to have economic significance. One system that has developed to define a discoverer in an interesting way is the system of claims, or prospecting. For my purposes the distinctive feature of a system of claims is that it permits ownership to occur prior to actual physical possession. In a system of claims, the discoverer is someone who has demonstrated a likelihood that the good exists, according to certain formal criteria.

There is an economic justification for such a system: it permits resources to be exploited more rationally. In many cases it is more efficient to delay before actually gaining physical possession of a resource, and such delay is possible only if ownership can be assigned prior to physical possession, otherwise a race would ensue to obtain the resource.

A somewhat fanciful example might help to make this point clearer. Suppose that several people enter a crowded auditorium at the same time, looking for seats. One outcome of this situation might be that the people race to whatever seats they spot, risking embarrassment and injury. Someone might lose several races, and be forced to run all around the auditorium before obtaining his seat.

A better outcome would result if there were a system for claiming seats. For example, the first person to spot a seat and inform the usher of its existence might have the right to that seat. Then the person would be able to walk to her seat in a dignified manner. When the two outcomes are compared, it can be seen that they have the same positive results: people occupying their seats. (Some people will arrive at their seats a bit later in the second case, since they walk rather than run; this is the desirable delay.) However, the first outcome has additional costs: injury, embarrassment, and wasted effort.

Yoram Barzel (1968) argues that resource-wasting races will occur in an economy that has a way of appropriating information (such as a patent system), as well as information in the public domain. The cost in his discussion is that resources are invested in developing the innovation earlier than would be the case optimally. While this cost is harder to grasp than the costs involved in racing for seats, it is no less real.

This general sort of problem can occur wherever there is common property that can be exploited for private purposes. In the case of minerals, the common property is the land (including the minerals buried in that land). In my auditorium example, the seats are common property; anyone may sit in them, until someone does, at which point the seat is "his" seat.

In the case of innovations, the common property is basic knowledge, which anyone can use. Thus, in Barzel's model, "innovations are assumed to be currently possible: we have the necessary blueprints and these are free. To make the innovation commercially feasible, however, a once-and-for-all investment is required" (Barzel 1968, p. 349). The blueprints represent basic knowledge and are common property. A critical point is that once the inventor makes the innovation commercially feasible, she can appropriate not only the value of that accomplishment (making the innovation commercially feasible), but also the value of the blueprints. It is the appropriation of (the value of) knowledge in the public domain that makes the comparison with other common property situations valid.

This is a complex point, so an example might be helpful. Suppose that a given piece of basic knowledge can serve as the foundation for only one worthwhile invention. In that case, once the invention has been patented, the piece of basic knowledge has no remaining (commercial) value, at least for seventeen years. One can calculate the value of the piece of basic knowledge—it is the difference between the cost of making the invention that is based on it and the market value of the patent on that invention. That value, which was in the public domain (anyone could have made the invention and captured it) is no longer in the public domain. It has been captured by the inventor. In the same way, the value of a mineral taken from public lands is captured by the one who takes it (assuming the taker is given ownership rights).

Edmund Kitch's theory of the patent system was, he says, inspired by Barzel's account. Kitch's specific contribution is to argue that the existing patent system takes the danger of races and other unnecessary costs into account. He suggests that many of its features should be understood as methods for preventing such races. These features make the patent system, in some respects, a prospecting system rather than a strict reward system. Thus, just as the minerals can be owned (as part of a claim) before they have all been dug up, so aspects of an IP good can be owned (under a patent) before they have been developed, according to Kitch.

In his view, intellectual property law (properly) gives ownership rights to more than what has been discovered at the time of patenting. The right that is given is to a *developed* invention rather than to the invention as the patentee has presented it for patenting. The patentee presents a working, but often quite crudely implemented, invention, yet in the end he owns the patent rights to the more sophisticated final product (Kitch 1977, p. 268).

A good that is yet to be discovered exists, and thus can be owned. On the other hand, a good that needs to be created does not yet exist, and so cannot be owned. By supposing that IP goods can be owned before they are discovered or created, Kitch is implicitly assuming that they are discovered goods.

Kitch shares with mainstream patent theorists the view that IP goods should be treated as rival rather than nonrival. Earlier in this discussion, I suggested that one motive for doing so is the fact that there are often a limited number of uses for an IP good. Kitch appears to have something like this sort of reasoning in mind. He argues that "information has appeared to be an example of something that can be used without limit. There is, however, a scarcity of resources that may be employed to use information, and it is that scarcity which generates the need for a system of property rights in information" (Kitch 1977, pp. 275–276, footnote omitted). I think that Kitch would fill his account out in terms of balancing scarce resources rather than talking of consuming an invention, but the end result is the same—treating IP goods in the same way that rival goods are treated, rather than as, for example, laws of nature are treated.

The articles in the bibliography that will be useful to the reader interested in this perspective include:

Barzel; Cheung; DeBrock; H. G. Johnson; Kitch (1977, 1980a); McFetridge and Smith; Yu.

In this volume: Dreyfuss; Winter.

In addition, many of the economic models cited in the section on ordinary property attempt to take into account Barzel's analysis.

IP Goods as Copyable Goods

The final mechanism I will consider is one that has only been proposed. There is no existing mechanism that explicitly deals with copyable goods as a category. Of course, if one says that IP goods do in fact fit into the copyable good category, then in a sense the intellectual property system as it exists is such a system. The sort of system I have in mind here is rather different. For example, in a 1944 essay, Michael Polanvyi proposes that a new system be developed, arguing that "in order that inventions may be used freely by all, we must relieve inventors of the necessity of earning their rewards commercially and must grant them instead the right to be rewarded from the public purse" (Polanvyi 1944, p. 65).

Thus, his system responds directly to the two features of copyable goods. First, because they are nonrival, and there is therefore no reason to restrict use, they are to be made available for all to use freely. Second, because they are created, the creator must be fully rewarded for his contribution.

The most natural objections to this sort of proposal are that it is impractical and unnecessary: impractical because of the difficulties in determining appropriate payment, and unnecessary because existing systems perform the job adequately. Objections could also be based on the grounds that IP goods are not in fact copyable goods, that is, created and nonrival. It might be argued, for example, that IP goods are discovered, not created, and thus that there is no such thing as a creator's right in IP goods that needs to be reflected in the system. This would still leave the need for appropriate incentives for discovery, however.

It might also be argued that IP goods are consumed in use on the grounds noted above. In this system there is no relationship between who benefits from an invention and who pays the reward, which would seem inappropriate for a good that is consumed. Polanvyi argues that "people in [a given] generation profit . . . about equally from the sum total of inventions made during their lifetime" (Polanvyi 1944, p. 65). This claim, if true, would resolve the problem, but the issue remains of interest if the claim is false (as it most likely is).

In fact, even if IP goods are not consumed in use, there remain some complex normative issues. For example, to judge the value of an invention, do we simply determine the savings it creates, or what people would have paid had payment been required? If the latter, do we take into account how much they *could* have paid? Polanvyi suggests that payment be based on information gathered from users, including "only data endorsable by accountants' certificate"

(Polanvyi 1944, p. 68). A full consideration of these issues is beyond the scope of this chapter, but they are critical to any such alternative to the intellectual property system.

Supposing that IP goods belong to the copyable goods category does not necessarily lead to proposing a new system to handle them. It can also lead to proposing that they be treated the same way as many other copyable goods, such as business procedures—that is, not protected at all. One justification for this approach is that such treatment is required if the competitive marketplace, which depends heavily on imitation, is to function properly.

The reader interested in either of these sorts of approaches might want to consider:

Allen; Baird; Benkay; Breyer; Gale; Hsia and Haun; Hurt; D. G. Johnson; Polanvyi; Tyerman; Wright.

In this volume: Dreyfuss; Kuflik.

Conclusion

A number of options were glossed over in my discussion. For example, it may be that all the goods handled by the intellectual property systems do not belong in the same category in Figure 15.1. It may be that copyrighted goods should be seen as created, while patented goods should be seen as discovered, or some other variation. Also, I did not discuss trade secrecy, which has a somewhat different structure from other forms of intellectual property, in that a trade secret necessarily has a limited number of users (otherwise, it is not a secret).

One should also consider the possibility that the categories discovered/created and nonrival/consumed-in-use do not apply cleanly to IP goods. IP goods (particularly in the case of patented goods) seem to have aspects of both discovered and created goods, depending on whether one focuses on the basic knowledge on which the good depends, or the creative effort involved in the making of that good. Similarly, when one focuses on the IP good itself, it seems to be nonrival, but when one considers the context in which it is used, it seems to have features in common with goods that are consumed in use. I have suggested that these facts help to explain the quite different stances that are taken with regard to the treatment of IP goods. To support that claim fully, a more fundamental analysis of these two distinctions would be necessary.

A Selected Annotated Bibliography of the Intellectual Property Literature

Patrick Croskery

Introduction

While the specifically ethical study of intellectual property does not yet have a literature of its own, there are many useful works, written by lawyers, economists, and philosophers in the course of their work on intellectual property law, the economics of innovation, and property theory, respectively. In addition, historians and sociologists occasionally consider issues that are of importance to someone interested in this topic.

The purpose of this bibliography is to give the interested reader access to these works. The bibliography is selected, because a full representation of the contexts in which the various articles and books are found would make the bibliography so large as to be useless. It is annotated so that a reader not familiar with the issues covered in each of the various contexts can still determine if a given article will be of interest.

The chapter that precedes this bibliography discusses at greater length the basis on which articles were selected, and also suggests a way of organizing these materials on normative grounds. In addition, in the course of that chapter the essays from this volume are associated with the materials in the bibliography. For these reasons, it might be helpful to consult Chapter 15 in order to make best use of the bibliography.

The following works can be found in this bibliography:

1. Discussions of *droit moral* (the moral right of authors)
2. Economic models and analysis of the patent and copyright systems
3. Historical discussions that include a normative dimension
4. A few articles on property theory (see Becker [1987] and Pennock and Chapman for bibliographies on this topic)
5. Discussions of the impact of new technologies, such as software and biotechnology, on the intellectual property systems
6. General discussions of the underlying justifications for the intellectual property systems
7. Reference works on this topic

8. Discussions of trade secrecy

9. Discussions of the interaction between the patent system and the university system

10. Discussion of the prospect theory of the patent system (the theory that the patent system resembles the system of claims that applies to minerals)

11. Alternative intellectual property schemes

12. Radical critiques of the existing systems

In order to maintain the focus of the bibliography, several issues that are related to intellectual property and have normative dimensions have been omitted. These include fair use and copyright law, antitrust and patent law, the Third World and patents, and software piracy. These issues, while potentially of interest to someone working in this area, are not nearly so central as the topics that were included.

At the end of the bibliography is a list of legal cases of particular interest, along with a brief description of the significance of each.

I would like to thank David Carey at the University of Pittsburgh for exchanging bibliographies.

The Bibliography

Adler, R. G. 1984. "Biotechnology as an Intellectual Property." *Science* 224: 357–363.
Offers a general discussion of the application of intellectual property law to biotechnology. Considers such factors as international competition and gene ownership.

Allen, R. C. 1983. "Collective Invention." *Journal of Economic Behavior and Organization* 4: 1–24.
Suggests that a fourth institution (in addition to the government/university, individuals, and individual firms) is responsible for some inventive activity. The institution, which he calls "collective invention," consists of otherwise competitive firms freely sharing advances. Suggests that this can occur where advances are difficult to conceal, and where advances are a by-product of ordinary business practices rather than specific R and D.

Arrow, K. 1962. "Economic Welfare and the Allocation of Resources for Invention." In *The Rate and Direction of Inventive Activity: Economic and Social Factors*, edited by R. R. Nelson, pp. 609–624. Princeton: Princeton University Press.
Analyzes the economics of inventive activity. Usually cited for the following conclusions: (1) "we expect a free enterprise economy to underinvest in invention and research (as compared with an ideal)"; and (2) "to the extent that a firm succeeds in engrossing the economic value of its inventive activity, there will be an underutilization of that information as compared with an ideal allocation" (p. 619).

Baird, D. G. 1983. "Common Law Intellectual Property and the Legacy of International News Service v. Associated Press." *University of Chicago Law Review* 50 (2): 411–429.

Argues that ideas and information should be presumptively free. Competition depends on imitation, and thus all innovation should not be appropriable. Includes a comparison of the natural-rights theory with the theory that intellectual property represents a balancing of incentives for the creator and public access.

Barzel, Y. 1968. "Optimal Timing of Innovations." *Review of Economics and Statistics* 50: 348–355.

Argues that innovations can be developed too early, relative to an optimal use of the resources involved in their development. This inefficient outcome will occur in economic systems that include both public-domain information and mechanisms for the appropriation of information (such as a patent system). Starting point for Kitch's (1977) analysis of the patent system.

Batzel, V. M. 1980. "Legal Monopoly in Liberal England: The Patent Controversy in the Mid-nineteenth Century." *Business History* 22 (2): 189–202.

Looks at the debate generated by abolitionists in the 1800s that led to some reforms at that time.

Beck, R. L. 1976. "Patents, Property Rights and Social Welfare: Search for a Restricted Optimum." *Southern Economic Journal* 43: 1045–1055.

Uses three case studies to show that innovators can respond to the structure of the patent system in ways that reduce social gains from innovation.

Becker, L. 1977. *Property Rights: Philosophic Foundations*. London: Routledge and Kegan Paul.

Offers a detailed analysis of the standard justifications for property rights, including arguments based on first occupancy, labor, utility, political liberty, and moral character. Sketches the foundations for a new theory of property based on this analysis.

______. 1987. "Property: A Select Bibliography." *Philosophy and Law Newsletter* (Winter 1987). Newark: University of Delaware.

Reviews the literature on property theory in a bibliographic essay.

Beier, F. 1980. "The Significance of the Patent System for Technical, Economic, and Social Progress." *International Review of Industrial Property and Copyright Law* 11 (5): 563–584.

Argues that the long history of patent systems demonstrates their effectiveness.

Beier, F., R. Crespi, and J. Straus. 1985. *Biotechnology and Patent Protection: An International Review*. Paris: Organisation for Economic Cooperation and Development.

Reports the results of survey responses by OECD member governments on issues related to patents and biotechnology. Issues covered include the adequacy of the patent system and user attitudes toward the patent system.

Bender, D. 1985/1986. "Protection of Computer Programs: The Copyright/Trade Secret Interface." *University of Pittsburgh Law Review* 47 (4): 907–958.

Argues that federal copyright law does not preempt state trade secret law. Considers the advantages of each form of protection and recommends using both kinds where possible.

Benkay, D. 1985. "The Patent Law of the People's Republic of China in Perspective." *UCLA Law Review* 33 (1): 331–378.

Offers a theoretical and historical analysis of China's new patent law, which confers a right to compensation, but not an exclusive right to exploitation (which would represent private ownership of the means of production and thus be inconsistent with socialism).

Bok, S. 1982a. "Secrecy and Openness in Science: Ethical Considerations." *Science, Technology, and Human Values* 7: 32–41.

Examines the specifically ethical problems that secrecy creates for scientists, particularly in an academic setting. Argues that the norm of openness should have greater force than such motives as profit seeking.

———. 1982b. *Secrets*. New York: Pantheon Books.

Includes discussions of trade secrecy (chap. 10) and secrecy in science (chap. 11). Relates these topics to the ethics of secrecy in general.

Branscomb, A. W. 1985. "Property Rights and Information." In *Information Technologies and Social Transformations*, edited by B. R. Guile, pp. 81–120. Washington: National Academy Press.

Analyzes information protection in terms of a set of (potentially conflicting) fundamental rights, such as the right to withhold information, the right to receive compensation for information, and so forth. Also provides a list of basic principles involved in the treatment of property rights in information, including the need for a public medium of transmission and the requirement that theft be punished.

———. 1986. "Who Owns Information?" Gannett Center for Media Studies, Occasional Paper No. 2. New York: Columbia University.

Discusses issues raised by the ownership of information. Includes a discussion of the ownership of names (in mailing lists), and the problem of piracy faced by the media. Contrasts "information priests" who desire the private ownership of information, and "information socialists" who want information to be free.

Breyer, S. 1970. "The Uneasy Case for Copyright." *Harvard Law Review* 84: 281–351.

Considers the effect of abandoning copyright. Believes that publishers would not be dramatically affected. Argues that harmful effects of copyright (higher cost,. lower distribution) are perhaps greater than the beneficial effects, and thus that copyright law should not be extended.

Brown, R. S. 1985. "Eligibility for Copyright Protection: A Search for Principled Standards." *Minnesota Law Review* 70 (2): 579–609.

Analyzes three principled approaches to copyright: author's rights, the constitutional balance between incentives and public access, and the economic view that public goods should be freely distributed. Suggests that all should be considered in creating law.

Bush, G. P., ed. 1972. *Technology and Copyright*. Mt. Airy, Md.: Lomond Systems.

Contains selected source materials on copyright issues. Includes a large annotated bibliography.

Canfield, J. 1984. "The Copyrightability of Object Code." *Notre Dame Lawyer* 59 (2): 412–430.

Contends that the resolution of the debate about whether object code should be copyrightable depends on the "philosophical" issues of who is an "author" and what is a "writing."

Carpenter, M. P., M. Cooper, and F. Narin. 1980. "Linkage between Basic Research Literature and Patents." *Research Management* 23 (2): 30–35.

Uses patent citations of scientific literature to show that basic, current research is very important to patentable innovations.

Chesser, J. 1985. "Semi-conductor Chip Protection: Changing Roles for Copyright and Competition." *Virginia Law Review* 71 (2): 249–295.

Argues that the chip law merely extends copyright law, generating uncertainty and creating boundary problems. Argues in favor of truly specific law for chips.

Cheung, S.N.S. 1982. "Property Rights in Trade Secrets." *Economic Inquiry* 20: 40–53.

Finds that trade secrecy has same economic problems as patents, only more so. Looks at conditions where secrecy is appropriate or historically entrenched.

Chisum, D. 1978 *Patents*. New York: M. Bender.

A standard reference work on patent law.

______. 1980. *Intellectual Property: Copyright, Patent, and Trademark*. New York: M. Bender.

A standard reference work on intellectual property law.

______. 1986. "The Patentability of Algorithms." *University of Pittsburgh Law Review* 47(4): 959–1022.

Argues that even mathematical algorithms should be patentable. Disagrees with the Gottshchalk v. Benson Supreme Court ruling on precedent and policy grounds.

Crawford, F. 1975. "Pre-constitutional Copyright Statutes." *Bulletin of the Copyright Society of the USA* 23 (1): 11–37.

Offers an analysis of the states' copyright statutes and their influence on the Constitution.

Dasgupta, P., R. Gilbert, and J. Siglitz. 1982. "Invention and Innovation under Alternative Market Structures: The Case of Natural Resources." *Review of Economic Studies* 49: 567–582.

Looks at innovation in various models: planned economy, pure monopoly, and competitive market with patents.

DaSilva, R. J. 1980. "*Droit Moral* and the Amoral Copyright: A Comparison of Artists' Rights in France and the U.S." *Bulletin of the Copyright Society of the USA* 28 (1): 1–58.

Discusses the moral right of an author to control his work, which plays a critical role in French copyright law, while economic reward plays the most important role in U.S. law.

Davidson, D. 1983. "Protecting Computer Software: A Comprehensive Analysis." *Jurimetrics* 23 (4): 339–425.

Focuses on the nature of software (tangible/intangible, utilitarian/symbolic) to determine the appropriateness of patent, copyright, and trade secret protection. Extensively reviews each type of protection. Concludes that copyright supplemented by trade secrecy is the most appropriate type of protection. Argues that the fact that software is created out of symbolic languages is what distinguishes it from other utilitarian works and makes copyright protection acceptable.

Davidson, D., and J. A. Davidson. 1986. *Advanced Legal Strategies for Buying and Selling Software: 1986 Cumulative Supplement* New York: John Wiley and Sons.
Considers current issues in software protection.

Davis, M. 1987. "Nozick's Argument for the Legitimacy of the Welfare State." *Ethics* 97: 576–594.
Uses a "Lockean" theory of property to justify the welfare state. Of particular interest is the status of "intangibles" (which include intellectual property) in that theory: "You produce an intangible insofar as what you do adds to the value of an already existing object without physically changing it" (p. 578). Considers objections to this account of intangibles.

DeBrock, L. M. 1985. "Market Structure, Innovation and Optimal Patent Life." *Journal of Law and Economics* 28: 223–244.
Constructs a model of patent system based on Kitch's prospect theory, rather than the standard reward theory.

Dorvee, S. M. 1981. "Protecting Trade Secrets through Copyright. " *Duke Law Journal* 1981 (6): 981–998.
Presents an analysis of a particular issue: the placing of copyright notices on trade secrets. The analysis is based on underlying purposes of the two forms of law. Argues that trade secrets are ideas, while copyright protects expression.

Dratler, J. 1979. "Incentives for *People*: Forgotten Purpose of the Patent System." *Harvard Journal on Legislation* 16 (1): 129–209.
Believes that the fact that the law allows employers to require employee-inventors to sign over rights to all potential inventions significantly reduces the incentive of inventors. Proposes dividing rights, based on a scale of "extraordinary effort" put forth by the employee—the greater the difficulty the inventor has, the greater her rights in the invention.

Dunn, S. A. 1986. "Defining the Scope of Copyright Protection for Computer Software." *Stanford Law Review* 38 (2): 497–534.
Discusses "substantial similarity" in the context of software copyrights. Proposes two new approaches: (1) protect the structure of a program as well as the code; (2) allow "unit registration" of screen displays and computer instructions, to simplify the process.

Dutton, H. 1984. *The Patent System and Inventive Activity during the Industrial Revolution, 1750–1852*. Manchester: Manchester University Press.
Discusses contemporary justifications and criticisms of the patent system during the Industrial Revolution. Considers the effect of that system on inventive activity. Suggests that the imperfect nature of the patent system might have accidentally resulted in the best outcome for technological progress.

Dworkin, G. 1981. "The Moral Right and English Copyright Law." *International Review of Industrial Property and Copyright Law* 12 (4): 476–492.
Reviews English law's equivalents for *droit moral* rights—finds that they are incompletely represented. Considers right to claim authorship, right to integrity, right of publication, and right of recall.

Elias, S. 1985. *Nolo's Intellectual Property Law Dictionary*. Berkeley: Nolo Press.
Presents the terms and concepts relevant to intellectual property law in a useful dictionary format. Intended for inventors, authors, and software writers.

Esezobor, J. E. 1976. "Concepts in Copyright Protection." *Bulletin of the Copyright Society of the USA* 23 (4): 258–267.
Considers four concepts behind copyright: economic (right to earn); personality (expression); privacy; public policy (means to end).

Evenson, R. E. 1983. "Intellectual Property Rights and Agribusiness R&D: Implications for the Public Agricultural Research System." *American Journal of Agricultural Economics* 65 (5): 967–975.
Argues for reorientation of public agricultural research toward basic science supportive of agribusiness. Offers data suggesting movement is in opposite direction.

Fellner, W. 1970. "Trends in the Activities Generating Technological Progress." *American Economic Review* 60: 1–29.
Discusses general issues involved in building a model of knowledge production.

Frederick, R., and M. Snoeyenbos. 1983. "Trade Secrets, Patents, and Morality." In *Business Ethics*, edited by M. Snoeyenbos, pp. 162–169. Buffalo: Prometheus Books.
Considers trade secrecy and patent issues in context of business ethics. Includes consideration of role of contracts.

French, W. J. 1971. "Scott Amendment to Patent Revision Act: Should Trade Secrets Receive Federal Protection?" *Wisconsin Law Review* 1971 (3): 900–921.
Discusses trade secrecy's boundary problems, its effect on employees, and so forth. Considers confidentiality versus property rights as grounding for trade secrecy.

Fudenberg, D., J. Stiglitz, and J. Tirole. 1983. "Preemption, Leapfrogging and Competition in Patent Races." *European Economic Review* 22: 3–31.
Uses an economic model to determine when patent races will be competitive—argues that possibility of leapfrogging, due to multistage R and D, is necessary for competition.

Fusfeld, H., and L. Peters, eds. 1982. *University-Industry Research Relationships*, NSB 82-2. Washington: National Science Board.
Contains studies dealing with all aspects of the university-industrial interface.

Gale, B. N. 1978. "Concept of Intellectual Property in People's Republic of China: Inventors and Inventions." *China Quarterly* 74: 334–355.
Examines the prize system, which was, at the time the article was written, the only system in China roughly equivalent to Western patent systems. Suggests that it serves three functions: to offer incentives; to aid in locating talented people; and to help organize information about inventions (a consequence of the registration process).

Gallins, N. T., and R. A. Winter. 1985. "Licensing in the Theory of Innovation." *Rand Journal of Economics* 16: 237–252.
Uses a model to determine the effect of licensing in different market settings. Finds that licensing encourages research when production costs of firms are similar, but discourages research when the costs are widely divergent. The role of patents in this model is not to limit the flow of information but, by giving property protection, to enhance it.

Goodman, J. S. 1984. "The Policy Implications of Granting Patent Protection to

Computer Software: An Economic Analysis." *Vanderbilt Law Review* 37 (1): 147–181.
Contends that economics of software industry demonstrate need for patent protection; the Supreme Court has misunderstood computer technology; and computer algorithms are not like laws of nature, but like processes, and should thus be given patent protection.

Grobstein, C. 1985. "Biotechnology and Open University Science." *Science, Technology, and Human Values* 10: 55–63.
Reviews the evidence concerning negative effects of industry influences on university research and finds it inconclusive. Suggests that, if negative effects are found, there are three responses: 1) ignore the problem, perhaps on the grounds that it is temporary; 2) have faculty use their powers of promotion and so forth to regulate faculty behavior; and 3) have outside forces, such as state governments in the case of state universities, impose regulations.

Harris, C., and J. Vickers. 1985. "Perfect Equilibrium in a Model of a Race." *Review of Economic Studies* 52: 193–209.
Offers a model of patent race that suggests that competition does not significantly affect winner's behavior. In effect, the race will be decided early on, based on R and D advantages or incentive differences.

Harvard Law Review. 1982. "Toward a Unified Theory of Copyright Infringement for an Advanced Technological Era." *Harvard Law Review* 96: 450–469.
Offers new categories of use to replace those found in current copyright law: iterative/interactive and commercial/noncommercial. Suggests that the law should allow free iterative, noncommercial use, on the grounds that technology will make anything else impractical.

———. 1983. "Copyright Protection of Computer Program Object Code." *Harvard Law Review* 96: 1723–1744.
Argues in favor of copyright protection, considering such issues as requirement of communication to humans, conflict with patent laws, and advantages over patent and trade secret protection.

Hirshleifer, J. 1971. "The Private and Social Value of Information and the Reward to Inventive Activity." *American Economic Review* 61: 561–574.

Honore, A. M. 1961. "Ownership," In *Oxford Essays in Juriprudence*, edited by A. G. Guest, pp. 107–147. Oxford: Clarendon Press.
Analyzes the concept of "ownership." Finds several features, such as the right to use and the right to destroy, to be critical to the concept. Suggests that the concept applies to copyright "in an extended and somewhat weaker sense than that in which it applies to material objects" (p. 131).

Horstmann, I., G. M. MacDonald, and A. Slivinski. 1985. "Patents as Information Transfer Mechanisms: To Patent or (Maybe) Not to Patent." *Journal of Political Economy* 93: 837–858.
Presents a model in which patents convey varying amounts of information beyond that contained in the patent (about possible alternative products) in order to explain why patenting activity is not strictly correlated with R and D expenditure. When a profitable alternative exists for competitors, firms will use trade secrecy rather than patents.

Hsia, T. T., and K. A. Haun. 1973. "Laws of People's Republic of China on Industrial and Intellectual Property." *Law and Contemporary Problems* 38 (2): 274–291.
Looks at China's intellectual property system. China did not at the time of this article have an intellectual property system comparable to those found in the West. The authors analyze this fact in the light of Marxist theory, and also attempt to explain the limited system of rewards that did exist.

Hull, D. 1985. "Openness and Secrecy in Science: Their Origins and Limitations." *Science, Technology, and Human Values* 10: 4–13.
Offers a functional analysis for the structure of scientific research activity. Suggests that it leads individual and social goals to coincide—the scientist seeks personal recognition, but must contribute to the goals of the institution to achieve it. Also considers historical factors.

Hurt, R. M. 1966. "The Economic Rationale of Copyright." In R. M. Schuchman, ed. *American Economic Review Papers and Proceedings* 56: 421–439.
Discusses philosophical theories (Locke, Kant) that might underlie copyright, and argues that copyright is not necessary or beneficial to society as a whole.

Idea: The Patent, Trademark, and Copyright Journal of Research and Education. Washington: George Washington University.
A journal that contains numerous articles of potential interest.

Johnson, D. G. 1985. "Should Computer Programs Be Owned?" *Metaphilosophy* 16: 276–288.
Considers the arguments for and against the ownership of computer programs. Distinguishes between ownership of the sort given by copyright and that given by patents. Concludes that current rights are morally appropriate.

Johnson, H. G. 1976. "Aspects of Patents and Licenses as Stimuli to Innovation." *Weltwirtschaftliches Archive* 112 (3): 417–428.
Discusses, from economic point of view, inefficiencies in the patent system: bias toward applied research, excessively rapid development, and so forth.

Journal of the Copyright Society of the USA. New York: New York University Law Center.
Another journal that contains articles of potential interest.

Journal of the Patent and Trademark Office Society. Arlington, Va.: Patent and Trademark Office Society.
Yet another journal that contains articles of potential interest.

Judd, K. L. 1985. "On the Performance of Patents." *Econometrica* 53: 567–585.
Constructs a model of the patent system using stream of innovations rather than single innovation. Analyzes the effect of different life-spans.

Kamien, M. I., and N. L. Schwartz. 1974. "Patent Life and R&D Rivalry." *American Economic Review* 64: 183–187.
Extends the Arrow-Nordhaus model of the patent system to conditions that include rivalry.

Katz, M. L., and C. Shapiro. 1985. "On the Licensing of Innovations." *Rand Journal of Economics* 16: 504–520.
Constructs a model using "perfect" patents to determine the incentives for licensing.

Finds that major innovations will not be licensed, although minor ones will be licensed under certain conditions.

Kayton, I. 1982. "Copyright in Living Genetically Engineered Works." *George Washington Law Review* 50: 191–218.

Argues, in a speculative fashion, that genetic creations are works that meet the criteria for copyright—the "tangible medium" being living matter in this case.

Kenny, M. 1986. *Biotechnology: The University-Industry Complex*. New Haven: Yale University Press.

Discusses the forms and effects of university-industry interactions in biotechnology. Argues that industry ties greatly affect the university's social ties and corresponding norms, most importantly that of openness.

Keplinger, M. S. 1981. "Computer Software: Its Nature and Protection." *Emory Law Journal* 30: 483–512.

Reviews the history of the Computer Software Copyright Act of 1980. Suggests that breaking software down into the process, the program, and the documentation helps in understanding the type of protection that is appropriate, whether it be patent, trade secret, or copyright protection.

Kidwell, J. A. 1985. "Software and Semiconductors: Why Are We Confused?" *Minnesota Law Review* 70: 533–577.

Suggests that software presents particularly difficult legal problems because it involves multiple levels of difficutly—problems arise at the level of facts, values, legal institutions, and language. Offers examples at each level.

Kitch, E. W. 1977. "The Nature and Function of the Patent System." *Journal of Law and Economics* 20: 265–290.

Argues that the patent system has, in addition to its reward function, a prospect function, similar to that of a mineral-rights system. This prospect function serves to prevent the inefficiently early development of innovations pointed out by Barzel (1968). In addition, the prospect function increases the overall efficiency with which innovations are developed, by placing control with one owner, thus avoiding redundant work.

———. 1980a. "Patents, Prospects, and Economic Surplus: A Reply." *Journal of Law and Economics* 23: 205–207.

Responds to McFetridge and Smith's (1980) comment on Kitch's (1977) article. Argues that the comment fails to take into account the role of transaction costs in Kitch's theory.

———. 1980b. "The Law and Economics of Rights in Valuable Information." *Journal of Legal Studies* 9 (4): 683–723.

Discusses the law and economics of information and firms, including trade secrecy but not patents. Argues that if trade secrecy is justified on the grounds that it creates incentives for the production of information, then by the same token it distorts that production, toward processes that are protectable by secrecy and away from those that cannot be kept concealed. Suggests that "property" is not the best framework for thinking about trade secrecy, because of the unique characteristics of "human capital."

Kitch, E. W., and H. Perlman. 1986. *Legal Regulation of the Competitive Process*.

Mineola, N.Y.: Foundation Press.
A standard casebook on intellectual property law.

Klitzke, R. A. 1986. "Trade Secrets: Important Quasi-property Rights." *Business Lawyer* 41 (2): 555–570.
A general consideration of trade secrecy law. Holds that trade secrets are better understood as a right of fairness than as a property right.

Kozinski, A. 1974. "Market Oriented Revision of Patent System." *UCLA Law Review* 21 (4): 1042–1080.
Proposes that the patent search aspect of Patent Office be privatized. Analyzes patent system as requiring speed, security, alienability, and notice.

Kwall, R. R. 1985. "Copyright and the Moral Right: Is an American Marriage Possible?" *Vanderbilt Law Review* 38 (1): 1–100.
Analyzes creator's moral right (versus copyright holder's economic interest). Discusses current limited protection of that right in U.S. law.

Lahore, J., G. Dworkin, and Y. Smyth. 1984. *Information Technology: The Challenge to Copyright*. London: Sweet and Maxwell.
Discusses, in the context of British law, several areas where technological advances have created difficulties for intellectual property law. Lahore contributes a discussion of reprographic reproduction; Smyth contributes one of audio and video recording, and another of broadcasting, cable, and satellite transmissions; and Dworkin contributes one on the nature of computer programs.

Lehmann, M. 1985. "The Theory of Property Rights and the Protection of Intellectual and Industrial Property." *International Review of Industrial Property and Copyright Law* 16 (5): 525–540.
Uses the history of property rights and economic theory to argue that patents are not harmful monopolies but rather, like ordinary property in modern economic theory, promote economic efficiency.

Levin, R. C. 1986. "A New Look at the Patent System." *American Economic Review* 76: 199–202.
Reflects on the survey discussed in Winter's article found in this volume. Suggests that the relative unimportance of patents in many industries has important public policy implications.

Lieberstein, S. 1979. *Who Owns What Is in Your Head?* New York: Dutton.
A practically oriented account of the laws of trade secrecy.

Lunn, J. 1985. "The Role of Property Rights and Market Power in Appropriating Innovative Output." *Journal of Legal Studies* 14 (2): 423–433.
Explains interindustry differences in R and D intensity by arguing that some industries are better able to appropriate innovations. Two features of innovations are particularly important: that they can be put into an observable product or process, as required by patent law; and that they meet the novelty requirement of patent law, which tends to favor more technologically advanced industries. Where patent protection does not operate, market power can aid in appropriating innovations.

McFetridge, D. G., and D. A. Smith. 1980. "Patents, Prospects, and Economic Surplus: A Comment." *Journal of Law and Economics* 23: 197–203.

Uses a model to examine the theory proposed by Kitch (1977) that the patent system might serve a prospect function. Concludes that "the prospect approach is not a useful framework within which to assess the merits of the patent system" (p. 203), because the apparent surplus generated by the prospect mechanism would be dissipated by a race for the patent.

Machlup, F. 1958. *An Economic Review of the Patent System: Study No. 15 of the Subcommittee on Patents, Trademarks, and Copyrights, Senate Committee on the Judiciary*. 85th Cong., 2d sess. Washington: Government Printing Office.

Analyzes the standard justifications for the patent system (the disclosure of secrets, the rewarding of inventors, and the offering of incentives) and finds that only the incentive theory has any plausibility. Concludes that if there were no patent system, it would be irresponsible to institute one, but since one exists, it would be irresponsible to abolish it. Draws this conclusion from the lack of information about the effectiveness of the patent system.

Machlup, F. and E. Penrose. 1950. "Patent Controversy in the Nineteenth Century." *Journal of Economic History* 10: 1–29.

Reviews the controversy over patents that arose in the nineteenth century. Uses the controversy to show that the same arguments for and against have been around a long time. Considers arguments related to the standard justifications of the patent system: that it represents a property (or natural) right, a just reward, an incentive to invent, or an incentive to make inventions public.

McMullin, E. 1985. "Openness and Secrecy in Science: Some Notes on Early History." *Science, Technology, and Human Values* 10:14–23.

Looks at the relation between openness and science in ancient Greece, during the Middle Ages, during the Renaissance, and up to the Statute of Monopolies (1623). Distinguishes between scientific secrecy and secrecy related to technology.

Mansfield, E. 1986. "Patents and Innovation: An Empirical Study." *Management Science* 32 (2): 173–181.

Presents the results of a survey indicating when inventions would or would not have been developed without the potential for patent protection. Finds that the patent system's effect on the rate of innovation is small.

Mansfield, E., M. Schwartz, and S. Wagner. 1981. "Imitation Costs and Patents: An Empirical Study." *Economic Journal* 91: 907–918.

Provides the results of a survey that asked if research would have been done if patents could not result. Half indicated not. Considers imitation costs even where patents are involved.

Mansfield, E., John Rapaport, Anthony Romeo, Samuel Wagner, and George Beardfly. 1977. "Social and Private Rates of Return from Industrial Innovations." *Quarterly Journal of Economics* 91: 221–240.

Presents an empirical study of economic effects of innovations. Considers social and private returns from nonrandom sample of firms.

Mark, J. J. 1977. "United States Copyright History and its Legislative History." *Law Library Journal* 70 (2): 121–152.

Offers a detailed legislative history of U.S. copyright law.

Massel, M. S. 1971. "Patent System and Economic Development." *New York University Law Review* 46 (3): 486–505.
Argues that patent system will be put under pressure by continuing innovation, environmental concerns, government involvement in the economy, and international relations. Suggests lines for research on the topic to handle these pressures.

Melman, S. 1958. *The Impact of the Patent System on Research: Study No. 11 of the Subcommittee on Patents, Trademarks, and Copyrights, Senate Committee on the Judiciary*. 85th Cong., 2d sess. Washington: Government Printing Office.
Explores the relationship between the patent system and nonprofit research. Finds that the patent system is a harmful influence, and in fact fails to fulfill its constitutional purpose of promoting science and the useful arts.

Milgrim, R. M. 1967. *Trade Secrets*, Vols. 12–12B of *Business Organizations*. New York: M. Bender.
A standard reference for trade secrecy law.

Miller, A., and M. Davis. 1984. *Intellectual Property: Patents, Trademarks and Copyright*. St. Paul, Minn.: West Publishing.
Offers a very straightforward and readable account of the relevant law. From the "In a Nutshell" series of legal guides.

Mislow, C. M. 1985. "Computer Microcode: Testing the Limits of Software Copyrightability." *Boston University Law Review* 65 (4): 735–805.
Uses the example of microcode (lowest level of software) to argue that all software should be copyrightable. Suggests that the difference between software and hardware is the presence of memory—this is what allows RAM and ROM but no circuits to be storage mediums, which can thus allow "fixed expression."

National Commission on New Technological Uses of Copyrighted Works (CONTU). 1979. *Final Report of the National Commission on New Technological Uses of Copyrighted Works*. Washington: Government Printing Office.
Contains the results and recommendations of the commission created in 1974 to study the effects of computers and photocopying on the copyright system. Recommends allowing the copyrighting of computer programs and data bases. These recommendations were followed.

Nelkin, D. 1984. *Science as Intellectual Property: Who Controls Scientific Research?* New York: Macmillan.
Suggests that the debates about secrecy in science and proprietary arrangements would be more productive if discussed in terms of a negotiated contract between science and society rather than in terms of "rights." Suggests that the participating parties must recognize the need to accommodate one another's goals.

Nimmer, M. B. 1975. "Photocopying and Record Piracy: Of Dred Scott and Alice in Wonderland." *UCLA Law Review* 22 (5): 1052–1065.
Considers fair-use photocopying as well as compulsory licenses and record duplication. Disagrees with a court ruling that "similar" does not allow duplication.

______. 1978. *Nimmer on Copyright: A Treatise on the Law of Literacy, Musical and Artistic Property, and the Protection of Ideas*. New York: M. Bender.
A standard reference work for copyright law.

———. 1985. *Cases and Materials on Copyright and Other Aspects of Entertainment Litigation*. St. Paul, Minn.: West Publishing.

A standard casebook on copyright law.

Nordemann, W. 1980. "A Right to Control or Merely to Payment? Towards a Logical Copyright System." *International Industrial Property and Copyright Law* 11 (1): 49–54.

Considers problems in the German system of compulsory licensing; for example, rapid technological changes create problems for the implementation of the "fair pay" requirement.

Nordhaus, W. D. 1969. *Invention Growth and Welfare*. Cambridge: MIT Press.

Offers a general economic model of invention, and in chap. 5, a model of a patent system, which he uses to derive the optimal life of a patent, which varies according to certain factors. Also compares patent and subsidy systems. Suggests that a patent system is best for "small" inventions, while a subsidy system is better for "large" (or "drastic") inventions.

Novos, I. E., and M. Waldman. 1984. "The Effects of Increased Copyright Protection: An Analytic Approach." *Journal of Political Economy* 92: 236–246.

Presents a model in which an increase in copyright protection decreases welfare loss due to underproduction but, contrary to expectations, does not increase welfare loss due to underutilization.

Office of Technology Assessment. 1986. *Intellectual Property Rights in an Age of Electronics and Information*, OTA-CIT-302. Washington: Government Printing Office.

Contains a collection of articles examining the past, present, and future status of intellectual property rights.

O'Hare, M. 1985. "Copyright: When Is Monopoly Efficient?" *Journal of Policy Analysis and Management* 4 (3): 407–418.

Argues that copyright is efficient only when certain conditions hold: a market for subsidiary uses, a high fixed cost of copying, and a market for copies.

Oppenländer, K. 1977. "Patent Policies and Technical Progress in the Federal Republic of Germany." *International Review of Industrial Property and Copyright Law* 8 (2): 97–122.

Offers an empirical study of the protective and informational influence of patents. A survey of firms finds that protection does help, particularly with uncertain R and D and innovations that are hard to keep secret. Finds the informational value of patents to be of limited use.

Panem, S. 1982. "The Interferon Dilemma: Secrecy versus Open Exchange." *Brookings Review* 1 (2): 18–22.

Examines in an informal manner the attitudes of researchers as a field moves from basic to applied (patentable) research.

Patterson, L. R. 1968. *Copyright in Historical Perspective*. Nashville: Vanderbilt University Press.

Views copyright history from a legal perspective. Argues that there are three ideas that have not received enough attention: "the danger of monopoly from the publisher rather than the author, the differing interests of the publisher and the author, and the rights of the individual user" (p. 228). Suggests that the author's creative interest should be recognized separately from her economic interest.

Pennock, J. R., and J. W. Chapman, eds. 1980. *Property: Nomos XXII*. New York: New York University Press.
Contains historical and analytical essays concerning the concept of "property" in general. Includes a selected bibliography on related topics.

Polanvyi, M. 1944. "Patent Reform: Monopolies for Pioneers." *Review of Economic Studies* 11: 61–76.
Proposes a radical reform of patent system. The government pays an inventor for his invention, and the invention is then free for others to use. The payment given is based on information gathered by the government from those who use the invention.

Raskind, L. J. 1986. "The Uncertain Case for Special Legislation Protecting Computer Software." *University of Pittsburgh Law Review* 47 (4): 1131–1184.
Suggests that the debate about possible sui generis protection of computer software need not be an either/or issue. Recommends congressional hearings on the issue.

Reich, L. S. 1977. "Research, Patents, and the Struggle to Control Radio: A Study of Big Business and the Uses of Industrial Research." *Business History Review* 51: 208–235.
Uses early radio history to argue that much research is undertaken for purposes that are unproductive from a social point of view, although they may be profitable for the companies involved.

Reinganum, J. F. 1982. "Dynamic Game of R&D: Patent Protection and Competitive Behavior." *Econometrica* 50: 671–688.
Uses a model to consider the effects of competition on innovation. Finds that where perfect patent protection is available, competition speeds up technical advance. In the case of imperfect patents, competition has conflicting effects.

Roberts, M. 1971. *Copyright: A Selected Bibliography*. Metuchen, N.J.: Scarecrow Press.
Offers an extensive bibliography on copyright issues.

Roeder, M. A. 1940. "Doctrine of Moral Right: A Study in the Law of Artists, Authors, and Creators." *Harvard Law Review* 53: 554–578.
Contains an early discussion of "moral right" (in contrast to economic right) of creators. These rights include the right to create or not to, paternity rights, and the right to prevent modifications.

Rosenzweig, R. M. 1985. "Research as Intellectual Property: Influences within the University." *Science, Technology, and Human Values* 10 (2): 41–48.
Looks at the development of industry-university relationships. Expresses concern about the impact of these relationships, as well as that of the patent system, on openness in the university environment. Followed by a commentary by H. W. Bremer, pp. 49–54.

Samuelson, P. 1984. "CONTU Revisited: The Case against Copyright Protection for Computer Programs in Machine Readable Form." *Duke Law Journal* 4 (1984): 663–769.
Argues that machine readable code should not be copyrightable because the disclosure requirement is not met, so that the work is not instructive; and utilitartian

works are not copyrightable. Suggests the creation of a specific new intellectual property law for machine readable code.

———. 1985. "Creating a New Kind of Intellectual Property: Applying the Lessons of the Chip Law to Computer Programs." *Minnesota Law Review* 70 (2): 471–531.
Argues that the same reasoning that led to the sui generis Chip Act leads to a sui generis law for machine readable programs, with an emphasis on the argument that copyright should not apply to utilitarian products. Also discusses patent/copyright "gap" and the place of the chip law in relation to it.

Scherer, F. M. 1972. "Nordhaus' Theory of Optimal Patent Life: A Geometric Reinterpretation." *American Economic Review* 62: 422–431.
Interprets Nordhaus's model using geometric methods. A reply by Nordhaus follows the article.

Schmid, A. A. 1985. "Biotechnology, Plant Variety Protection, and Changing Property Institutions in Agriculture." *North Central Journal of Agricultural Economics* 7: 129–138.
Considers the effects that applying the concept of property to plants has on research. Suggests that the particular character of plants presents several difficulties. First, property protection results in a bias toward creating unnecessarily stable varieties. In addition, the property right itself is narrow because firms can often make merely cosmetic changes and create a new plant. This second difficulty results in a bias against the rise of genetic technology where such changes are possible, and at the same time a needless diversity of plant varieties.

Shapiro, C. 1985. "Patent Licensing and R&D Rivalry." *American Economic Review* 75 (May): 25–30.
Analyzes the effects of licensing and imitation on diffusion of inventions and the incentives for their creation.

Sherwood, M. 1983. "The Origins and Development of the American Patent System." *American Scientist* 71 (5): 500–506.
Examines the history of the U.S. patent system with particular attention to the relationship of patents to "the goals of a democratic society" (p. 500).

Soltysinki, S. J. 1986. "Are Trade Secrets Property?" *International Review of Industrial Property and Copyright Law* 17 (3): 331–356.
Argues for the formal recognition of trade secrets as property. Suggests that this would clear up needless confusion in business contexts where the concept "property" is used—for example, in securities law. Holds that trade secrets are already being used as property, just not consistently, and that "property" as a legal category is broad enough to include trade secrets meaningfully.

Stedman, J. C. 1947. "Invention and Public Policy." *Law and Contemporary Problems* 12 (4): 649–679.
Considers the standard justifications and criticisms of the patent system. Concludes that there are several significant problems with the patent system, including the complexity of the legal procedures, and the lack of discrimination reflected in the seventeen-year grant, among others.

Stevenson, R. B., Jr. 1980. *Corporation and Information: Secrecy, Access, and Disclosure*. Baltimore: Johns Hopkins University Press.

Considers the flow of information within and from corporations. Chap. 2 looks at trade secrecy, and concludes that the law is "a clumsy instrument for achieving its avowed ends" (p. 25).

Strauss, W. 1960. *The Moral Right of the Author: Study No. 4 of the Subcommittee on Patents, Trademarks, and Copyrights, Senate Committee on the Judiciary*. 86th Cong., 1st sess. Washington: Government Printing Office.
Argues that U.S. law gives as effective protection to the personal rights of authors as that given abroad under the heading "moral right of the author."

Tandon, P. 1982. "Optimal Patents with Compulsory Licensing." *Journal of Political Economy* 90: 470–486.
Uses a model to argue that compulsory licensing can lead to social welfare gains. Uses optimal rather than "reasonable" royalty rate.

Tyerman, B. W. 1971. "Economic Rationale for Copyright Protection for Published Books: Reply to Professor Breyer." *Bulletin of the Copyright Society of the USA* 19 (2): 99–128.
Responds in a point-by-point fashion to Breyer's (1970) criticisms of the copyright system.

Valentin, J. B. 1975. "Copyright—Moral Right: Proposal." *Fordham Law Review* 43 (5): 793–819.
Argues for the moral right of authors from historical/conceptual angle, just prior to 1976 revision of the Copyright Act.

Von Hippel, E. 1982. "Appropriability of Innovation Benefit as a Predictor of the Source of Innovation." *Research Policy* 11: 95–115.
Notes the fact that in some industries product users and in others the manufacturers are usually the source of innovations. Argues that this can be explained by the ability of the user or manufacturer to appropriate the benefit. Users appropriate value by forming an industrywide quasi monopoly, while manufacturers do so through patents, trade secrecy, or the lead-time advantage. Argues that patents are not very effective.

Webster, J. O. 1984. "Copyright Protection of Systems Control Software Stored in ROM Chips: Into the World of Gulliver's Travels." *Buffalo Law Review* 33 (1): 193–224.
Opposes the Apple v. Franklin ruling on the grounds that copyright must involve expression to persons.

Weiner, C. 1986. "Universities, Professors, and Patents: A Continuing Controversy." *Technology Review* 89 (2): 33–43.
Discusses the issues raised in his article in this volume in a popular journal.

Wincor, R. 1971. "Beyond Copyright." *Bulletin of the Copyright Society of the USA* 19 (1): 48–56.
Suggests five new categories to replace copyright: original works, emblems, performances, secrets, and transmissions.

Woodward, W. R. 1942. "Reconsideration of the Patent System as a Problem of Administrative Law." *Harvard Law Review* 55: 950–977.
Examines four problems related to the patent system: 1) private parties will often not challenge invalid patents because of the legal expenses; 2) the patent system does not distinguish levels of invention—it is an all-or-nothing process; 3) a plaintiff recovers his

legal costs in damages if he wins, whereas a successful defendent merely gets a dismissal of the claim, so costs are borne disproportionately; and 4) court processes can delay the expiration of the monopoly to the detriment of the public interest.

Wright, B. D. 1983. "The Economics of Invention Incentives: Patents, Prizes, and Research Contracts." *American Economic Review* 73: 691–707.
Analyzes the advantages and disadvantages of patents, prizes, and contracts under differing circumstances.

Yu, B. T. 1984. "A Contractual Remedy to Premature Innovation: The Vertical Integration of Brand Name Specific Research." *Economic Inquiry* 22: 660–667.
Suggests that vertical integration of research can discourage premature innovation of the sort considered by Barzel (1968) and Kitch (1977).

Cases cited in the course of this volume which are of particular interest:

Apple v. Franklin (1983), which permitted the copyrighting of object code in ROM

Diamond v. Chakrabarty (1980), which permitted the patenting of life-forms

Diamond v. Diehr (1981), which permitted the patenting of a process that included software

Gottschalk v. Benson (1972), which denied a patent to a piece of software on the grounds that protection of an algorithm would be too broad

Hubco Data Products v. Management Assistance, Inc. (1983), which held that the making of copies of a program in the course of reverse engineering it represented copyright infringement

Parker v. Flook (1978), which denied a patent to a piece of software on grounds similar to Gottschalk v. Benson, but prepared the way for Diamond v. Diehr

Additional cases can be found in relevant casebooks: Chisum (1978, 1980); Kitch and Perlman (1986); Milgrim (1967); and Nimmer (1975).

Notes on Contributors

JILL BELSKY is a Ph.D. candidate in the field of Development Sociology at Cornell University. Her work concentrates on the relationship between technological change and environmental quality In Southeast Asian agriculture.

LEONARD G. BOONIN is Professor of Philosophy at the University of Colorado in Boulder. His research interests are in philosophy of law, ethics, and social policy. He is associated with the Center for Values and Social Policy at the University of Colorado.

FREDERICK H. BUTTEL is Professor of Rural Sociology at Cornell University. His research interests are in technological and social change, particularly in relation to agricultural research, biotechnology, and international development.

PATRICK CROSKERY is working on a Ph.D. in philosophy at the University of Chicago. He is writing his dissertation on the foundations of justice, paying particular attention to issues related to intellectual property.

CARL CRANOR is Professor and Chairman of Philosophy at the University of California, Riverside. He works on legal and moral philosophy with an emphasis on laws regulating science and technology. Recent studies include research into federal regulation of toxic substances.

DUNCAN M. DAVIDSON is a lawyer and a consultant with The MAC Group in San Francisco. He specializes in strategies within the information industry. Recent projects include an assessment of the impact of new technologies on the data-base software market.

MICHAEL DAVIS is a philosopher and Senior Research Associate at the Center for the Study of Ethics in the Professions at the Illinois Institute of Technology. His many publications on the philosophy of law and ethics include a study of a Lockean theory of property (*Ethics*, 1987).

ROCHELLE COOPER DREYFUSS is Professor of Law at New York University School of Law. She specializes in problems posed by the interaction of science and law, especially the protection of information through patent, copyright, trademark, and related legal regimes.

GERALD DWORKIN is Professor of Philosophy at the University of Illinois at Chicago. His many works in ethical theory include *The Theory and Practice of Autonomy*, recently published by Cambridge University Press.

ALAN GOLDMAN is Professor and Chairman of Philosophy at the University of Miami. His extensive publications in ethical theory include the influential volume, *The Moral Foundations of Professional Ethics*.

PATRICK KELLY is an engineer and lawyer. He works at Haverstock, Garrett and Roberts, a patent law firm in St. Louis specializing in biochemistry, genetic engineering, and medical technology. The author of four books, he holds one issued patent and has several pending applications.

ARTHUR KUFLIK is Associate Professor of Philosophy at the University of Vermont. He specializes in moral and social philosophy and has written on the theory of economic justice, the inalienability of autonomy, inalienable rights, and commonsense morality.

ALAN J. LEMIN is President of Pharmaceutical Licensing Consultants Company in Shell Beach, California. He was formerly a research scientist and Director of Research Contracts at Upjohn Company, where he was responsible for much basic research technology acquisition, particularly of pharmaceuticals, biotechnology, and agriculture.

PAMELA SAMUELSON is Professor of Law at the University of Pittsburgh. She has written extensively on software intellectual property rights. During the 1988–1989 academic year she taught computer law at the Emory University School of Law.

JOHN W. SNAPPER is Associate Professor of Philosophy and Associate Dean of Lewis College of Sciences and Letters at the Illinois Institute of Technology. He has worked extensively on social, legal, and ethical issues that affect the computer and electronics industries.

VIVIAN WEIL is Director of the Center for the Study of the Ethics in the Professions and Associate Professor of Ethics at the Illinois Institute of Technology. She works on issues of ethics and social responsibility in engineering, science, and business, with a recent focus on controls on the flow of scientific information.

CHARLES WEINER is Professor of History of Science and Technology in the Program in Science, Technology, and Society at the Massachusetts Institute of Technology. His research focuses on political and social dimensions of contemporary science, with a special emphasis on the public controversies over biotechnology and academic patenting.

SIDNEY WINTER is Professor of Economics and Management at the Yale School of Organization and Management. His research interests are in the areas of business firm behavior, technology, and competitive strategy.

Index